FIVE EASY STEPS

To a Balanced Science Program

for Primary Grades

Lynn F. Howard

Kindergarten to Grade 2

LEAD+
LEARN
PRESS

ENGLEWOOD, COLORADO

The Leadership and Learning Center
317 Inverness Way South, Suite 150
Englewood, CO 80112
Phone 866.399.6019   Fax 303.504.9417
www.LeadandLearn.com

Published by Lead + Learn Press, a division of Advanced Learning Centers, Inc.

Library of Congress Cataloging-in-Publication Data

Howard, Lynn F., 1954-
    Five easy steps to a balanced science program for primary grades / Lynn F. Howard.
        p. cm.
    ISBN 978-1-933196-95-4 (alk. paper)
    1. Science--Study and teaching (Primary)--United States. I. Title.

LB1532.H75 2010
372.3'5--dc22

2010036233

ISBN 978-1-933196-95-4

Printed in the United States of America

14 13 12 11 10   01 02 03 04 05 06 07 08

Contents

Introduction

The *Five Easy Steps*

CHAPTER 1

Step 1: Establishing an Effective Science Environment

CHAPTER 2

Step 2: Problem Solving

CHAPTER 3
Step 3: Conceptual Understanding

CHAPTER 4
Step 4: Mastery of Science Information

CHAPTER 5

Step 5: Common Formative Assessments

Inside the Classroom

CHAPTER 6

Inside the Kindergarten Classroom

CHAPTER 7

Inside the First-Grade Classroom

CHAPTER 8

Inside the Second-Grade Classroom

PART THREE Resources for Implementation

CHAPTER 9

Putting It All Together

APPENDIX

Reproducibles

Acknowledgments

Many individuals have contributed to the development of the books in the *Five Easy Steps to a Balanced Science Program* series.

Larry Ainsworth and Jan Christinson, authors of *Five Easy Steps to a Balanced Math Program*, inspired me to write *Five Easy Steps to a Balanced Science Program*. They provided a solid, research-based model for the development of these resources, and I will always be indebted to both of them for their encouragement and support.

To the many elementary, middle, and high school educators who have implemented the *Five Easy Steps* and have made suggestions, I am eternally indebted.

A special thank-you goes to:

Wallace Howard, my husband, who patiently endured hours and hours of endless typing and who taught me that "if it doesn't work in kindergarten, it just won't work."

My parents, who gave me independence, a love of investigation, and curiosity.

Catherine Koontz, my former student and now colleague, who inspired me to write *Ready for Anything: Supporting New Teachers for Success* and to follow my dream and love of science. And to her second-grade students, who just love experimenting at Mineral Springs Elementary School in Winston–Salem, North Carolina.

Elizabeth Daly, my friend and colleague, who offered continual support and encouragement.

McKinley Johnson and John Fout, former first-year, now veteran, science teachers, who taught me to never, ever forget what it was like to be a first-year science teacher. And to Mr. Johnson's eighth-grade students at Lake Norman Charter School in Huntersville, North Carolina.

Ray Brawley, my friend and science mentor, who told me how many days I had until retirement and who taught next door to me for eighteen years.

Jennifer Seabourne and her fifth-grade students at Mineral Springs Elementary School in Winston–Salem, North Carolina.

Michelle Byrd and her kindergarten scientists at Long Creek Elementary School in Huntersville, North Carolina.

And in memory of Tommy Tucker, my high school biology teacher, who always believed in me and who instilled in me an intense love of science.

About the Author

 Lynn Howard is a Professional Development Associate for The Leadership and Learning Center. She worked in the Charlotte–Mecklenburg School System for more than thirty years as a middle grades science teacher, coordinator of the gifted program, and a Regional Assistant Superintendent for middle schools. Lynn has conducted extensive staff development around the country and has brought a wealth of experience to schools and districts, including improving planning and teaming, leadership and relationship building, and systemic school-reform strategies. As a presenter on a National Speaker's Bureau for Geoscience Education, Lynn has provided seminars and workshops on best instructional practices and effective site-based teacher-retention programs. And as someone who has survived being locked in an outhouse miles from civilization, she brings an adventurous approach to the real-life strategies for teaching and learning.

Lynn has been invited to present at conferences of the Association of Supervision and Curriculum Development, the National Science Teachers Association, Phi Delta Kappa, the National Staff Development Council, the International Reading Association, and other state and regional conferences. Her philosophy is to make a difference in education and totally involve the audience with the presentation. She has a passion for teaching, and it is reflected through her enthusiasm and commitment to high-quality, professional development. Whether working with teachers or principals, Lynn is always ready to share her excitement for new opportunities and adventures.

One of her middle school principals says, "Lynn is always welcome at my school because she brings a realistic and teacher-friendly approach to professional development. My teachers know that they will take something away that works and can be used the next day in the classroom."

Lynn holds a Master's Degree in Earth Science and Education and a Bachelor's Degree in Biology from the University of North Carolina at Charlotte, Academy Certification from the National Staff Development Council, and National Certification in Gifted and Talented.

Ready for Anything: Supporting New Teachers for Success is Lynn's first book. With a passion for helping first-year teachers, Lynn is an advocate for those entering the teaching profession.

Lynn lives with her husband, Wallace, in Huntersville, North Carolina. Lynn's husband, who taught kindergarten for 26 years, is her sounding board and reality check. They are always talking about new teaching ideas. They both enjoy gardening and traveling and just recently visited Alaska, their fiftieth state.

Preface

As long as I can remember, I was always fascinated with science. As a child, my parents never discouraged me in the explorations around the house and yard. My mother was very patient when I would bring in dead frogs or creepy-crawly things. And my dad was oblivious to my sister and me collecting rocks in Maine, until he picked up his golf bag. I was able to spend hours on the coast of Maine, exploring rocks and tide pools and the smelly things that hid under the seaweed. I think that my ultimate love of science blossomed when I saw my grandmother give mouth-to-mouth resuscitation to a frog. She had accidentally dropped the garden hose on him and had knocked the wind out of him. My grandmother, the frog, and I all survived the ordeal.

I was fortunate to participate in the local and state science fairs during middle school. Again, my parents allowed me to pursue my love of inquiry but did not count on the number of mice and the horrific smell from three years of research. My mother did not see the humor in this project after several of the female mice gave birth on the way to the state fair in Melbourne, Florida. And after winning several awards, the entire house and the draperies had to be cleaned and disinfected once the project was over. I don't think they will ever forgive me for this project, but they did encourage me to move forward with my science education.

My high school biology teacher, Tommy Tucker, instilled in me the love of the marine world and would take us on weeklong field trips to Crystal River, Florida. There we would spend a week at a marine science station doing projects, exploring the ocean, and learning how to relate to the natural world. We were able to swim with the manatees, catch crabs for dinner, and throw nets, just to see what lived out in the waters.

My goal in life was to be a pathologist. My mother and step-grandmother were teachers, and I had no use for this profession. However, plans changed when I worked at a juvenile delinquency center in Atlanta, Georgia. I was the recreation director and tutor for thirteen young men incarcerated in a detention center. I fell in love with teaching, and because I had no desire to be a language arts teacher, I chose my first love—science. My college English composition teacher suggested this field because I had no talent for writing. Although I did my student teaching in advance-placement chemistry and biology, I found myself teaching eighth-grade earth science. Because I had no earth science background, I felt that I was just one day ahead of my students. For eighteen years, I taught earth science and oceanography, along with math and reading.

Throughout my teaching career, I was involved with professional development and service to science organizations. I had the opportunity to serve as a director with the North Carolina's Science Teachers Association, which made me realize the need for ongoing support for science education.

It was important to me to instill the passion for science in my students. Having the opportunity to visit all fifty states made it possible for me to design "in-class" field trips for the kids. As I traveled the United States, I wrote and created integrated units that included history, art, music, literature, and the science of the area. We went everywhere (in class), and my students still remember the fun of "On the Road" and "Where Am I?"

My mother helped perpetuate my passion for teaching science. In an excited voice, she called me from the mountains one day and told me that she had captured a large copperhead snake that was giving birth. She placed it in a jar, covered it with alcohol, and kept it for me. I had that snake in my classroom for years, and the kids loved being able to observe this fascinating creature.

Although *Five Easy Steps* was not published during those years, I followed these steps in my own classroom. When state testing and accountability arrived in North Carolina, all of us were concerned that we could not "teach our favorite units" but would be mandated to follow the state curriculum. Even in the turbulent years that followed, I still found that the steps were timeless and matched all of the best instructional practices.

As the years went by, I was fortunate to have been given many awards for teaching science, including local, state, and national recognition. One my favorite awards was for the North Carolina and Southeast Outstanding Earth Science Teacher. After receiving the award, I wanted to give something back to the National Association of Geoscience Teachers and agreed to serve on its executive board and on the National Speaker's Bureau. These were wonderful opportunities to share my enthusiasm and experience in teaching science.

For those of us in the field of science, it is imperative that we support and encourage new science educators. Elizabeth Daly was one of my student teachers who sat in front of me at one of the award ceremonies and became a stellar ninth-grade physical science teacher. As a Regional Assistant Superintendent, I had the opportunity to work with numerous first-year teachers. Many of them had no formal science or education training, and support was critical for their success. John Fout and McKinley Johnson were classic examples of young educators who had the desire to teach but just needed the added help and guidance. They constantly remind me of the importance of ongoing professional development for quality science teaching and learning.

SCIENCE is just plain FUN! This book is designed to help elementary science teachers plan for an adventurous and fulfilling year of intrigue, mystery, and experiments. Pick and choose the activities, adapt and change, and do your own thing, but get kids excited about science at the beginning.

One of the best things that you can do for your students is to perpetuate their sense of wonder and willingness to experiment. Encourage them to question, to invent, to explore, to wonder, and to understand that making mistakes takes courage and is necessary for growth.

Open the door for them in the first few days, continue to expand and challenge them with problems, and encourage all students to be thorough and complete in their investigations.

Good luck, and best wishes for a successful science year!

Introduction

Essential Question

How can a busy classroom teacher build science-confident students who understand the content, solve problems, communicate their knowledge, and perform well on standardized achievement tests?

✎ KEY POINTS

1. You need to understand the strengths, weaknesses, and challenges of your science program, which will foster growth and improvement.

2. The National Science Education Standards are designed to guide the nation toward a scientifically literate society.

SCIENCE EDUCATION TODAY

HOW DO CHILDREN LEARN ABOUT SCIENCE and how to do science in today's world? Book One of *Five Easy Steps to a Balanced Science Program* provides a framework for guiding science educators to help students achieve scientific literacy.

The question is how science educators can sustain this sense of curiosity over time. The opportunity to teach children how to learn about science is tremendous, and we must take advantage of the natural inquisitiveness that young children have about the natural world. We must also be aware of the teaching and learning resources and methods that best address quality science instruction. And last, we must provide teachers with professional development that allows them to learn how to teach science with access to the most current resources and materials and adequate instructional time.

This country established a goal through the development of the National Science Education Standards that all students should achieve scientific literacy. This is the vision of science education that will make science literacy a reality in the twenty-first century and beyond. Primary educators set the stage for the beginning learner by making it possible for them to explore and learn about the world around them. A sound background and understanding of science concepts and principles will give children the skills and knowledge needed to solve problems, think creatively, work cooperatively, and transfer their learning to real-world situations.

The primary teacher is responsible for teaching everything in all content areas. Children must be taught management and how to work together. They must learn procedures, rules, and social skills. In the primary grades, science encompasses a variety of topics that cover a wide range of content, including weather, animals and plants, the environment, and the human body. Teachers do not need to be experts in all of these areas but must have a basic understanding of all of the content represented in their curriculum standards. They also must have a working knowledge of best instructional practices so that students are allowed to develop interpersonal and intrapersonal skills.

So I welcome you to the *Five Easy Steps to a Balanced Science Program*. In this time of standards, assessments, and accountability, try to remember why you chose to become a teacher. The love of working with children goes beyond the textbook and curriculum and reaches into the world of enthusiasm and the sense of wonder about the things around you.

The design of *Five Easy Steps to a Balanced Science Program* is supported by the National Science Education Standards, which outline what students need to know, understand, and be able to do in order to be scientifically literate at various grade levels. The following section, "The National Science Education Standards," has an overview of the eight categories of the content standards.

The National Science Education Standards*

The National Science Education Standards are designed to guide our nation toward a scientifically literate society. Founded in exemplary practice and research, the Standards describe a vision of the scientifically literate person and present criteria for science education that will allow that vision to become reality. The goals for school science that underlie the National Science Education Standards are to educate students who are able to:

- Experience the richness and excitement of knowing about and understanding the natural world

- Use appropriate scientific processes and principles in making personal decisions

- Engage intelligently in public discourse and debate about matters of scientific and technological concern

- Increase their economic productivity through the use of the knowledge, understanding, and skills of the scientifically literate person in their careers

Schools that implement the standards will have students learning science by actively engaging in inquiries that are interesting and important to them. Students will establish a knowledge base for understanding science. Teachers will be empowered to make decisions about how students learn, how they learn it, and how resources are allocated.

Setting national goals and developing national standards to meet them are recent strategies in our education reform policy. Support for national education standards by state governments originated in 1989, when the National Governors Association endorsed national education goals. The first standards appeared in 1989, when mathematics educators and mathematicians addressed the subject of national standards with two publications: *Curriculum and Evaluation Standards for School Mathematics* by the National Council of Mathematics (1989) and *Everybody Counts: A Report to the Nation on the Future of Mathematics Education* by the National Research Council (1989).

In the spring of 1991, the National Science Teachers Association president asked the National Resource Center to coordinate development of National Science Education Standards. Major funding for this project was provided by the Department of Education and the National Science Foundation.

After many suggestions for improving the predraft were collected and analyzed, a revised standards document was prepared as a public document and released for nationwide review in December 1994. More than 40,000 copies of

---

*Reprinted with permission from *National Science Education Standards*, 1996, by the National Academy of Sciences. Courtesy of the National Academies Press, Washington, D.C., pp. 11, 13, 14, 15, 104, 105, 106.

the draft National Science Education Standards were distributed to 18,000 individuals and 250 groups. The comments of the many individuals and groups who reviewed the draft were collaged and analyzed, thus preparing the final copy of the National Science Education Standards. The National Science Education Standards outline what students should know, understand, and be able to do in natural science. The content standards are a complete set of outcomes for students; they do not prescribe a curriculum. The standards were designed and developed as a component of the comprehensive version of the final document.

There are eight categories of content standards, each with integrative conceptual and procedural schemes:

1. Unifying concepts and processes in science

The standards describe some of the integrative schemes that can bring together students' experiences in science education. These concepts and processes can be the focus of instruction at any grade level, but should always be closely linked to outcomes aligned with other content standards:

- Systems, order, and organization
- Evidence, models, and explanation
- Change, consistency, and measurement
- Evolution and equilibrium
- Form and function

2. Science as inquiry

The standards on inquiry highlight the ability to conduct inquiry and develop understanding about scientific inquiry. These standards approach science through skills such as observation, inference, and experimentation. Students are required to combine science processes with scientific knowledge as they use scientific reasoning and critical thinking. Science as inquiry is basic to science education and is a controlling principle in the organization and selection of students' activities and scientific work.

- Understanding of scientific concepts
- An appreciation of "how we know" what we know in science
- Understanding of the nature of science
- Skills necessary to become independent inquirers about the natural world
- The disposition to use the skills, abilities, and attitudes associated with science

3. Physical Science, 4. Life Science, and 5. Earth and Space Science Standards

The standards for physical, life, earth, and space science describe the subject matter of science using three widely accepted divisions of science. The subject matter focuses on the scientific facts, concepts, principles, theories, and models that are important for all students to know, understand, and use.

6. Science and Technology

The Science and Technology Standards establish connections between the natural and designed worlds and provide students with opportunities to develop decision-making skills. These standards emphasize abilities associated with the process of design and fundamental understandings about science and the connection to technology.

7. Science in personal and social perspectives

These standards address the need for students to understand and act on personal and social issues. The content in these standards provides a foundation on which to base decisions as citizens.

8. History and nature of science

The history and nature of science standards allow students to understand and reflect on the theme that science is ongoing and changing. The standards recommend the use of history in science to clarify scientific inquiry, the human component of science, and the role that science has played in the development of different cultures.

THE PURPOSE OF *FIVE EASY STEPS*

Five Easy Steps to a Balanced Science Program is designed to build confidence and competency in science educators so that student achievement is impacted. The National Science Teachers Association (NSTA) states: "To be prepared for the twenty-first century, it is critical that all students have sufficient knowledge of and skills in science. Studies suggest that high-quality teaching can make a significant difference in student learning. To achieve this goal, schools and school systems must devote time and resources to effective professional development for all K–12 teachers of science and science educators to support learning throughout their careers" (NSTA Position Statement, May 2006).

When students participate in a "balance of science" activity, they are engaging in learning that will:

- Build competency and confidence in their science ability

- Develop scientific reasoning and problem-solving abilities

- Deepen conceptual understanding

- Allow them to demonstrate their understanding in a variety of assessment formats

Teachers who read *Five Easy Steps to a Balanced Science Program* will:

- Learn how to implement the five steps of a balanced science program

- Learn how to establish an effective science classroom environment

- Understand science problem solving and the science process skills

- Learn the components of a conceptual science unit

- Learn and apply best instructional strategies for teaching science and integrating literacy into the science content

- Discuss components of science lessons that research shows have an impact on student achievement

- Learn specific strategies and activities that improve student learning in science

- Integrate the *Five Easy Steps* with existing science-adopted programs and state and local standards

DESIGNING THE BALANCED SCIENCE PROGRAM

Five Easy Steps to a Balanced Science Program is modeled after *Five Easy Steps to a Balanced Math Program* by Larry Ainsworth and Jan Christinson. Although there are similarities in the format, science lends itself to additional topics and content-specific organization, management, and instructional strategies that should be implemented in science classrooms. The design of *Five Easy Steps to a Balanced Science Program* maintains the fidelity of the math series but includes best practices for implementing effective science teaching and learning at all grade levels.

The *Five Easy Steps to a Balanced Science Program* series provides a model for immediate and ongoing improvement in science instruction in grades K–12. When students are engaged in a "balance" of science instruction and activities, they can experience success in applying their science skills and reasoning ability to solve real-life problems that require scientific thought processes. *Five Easy Steps* gives classroom teachers a series of five steps that provide a "balance" for their science instruction and assessment. The goal of this process is to provide opportunities for elementary, middle school, and high school science educators to develop the understanding and skills that are necessary for impacting science learning and achievement among students.

Each of the three books published under the series title *Five Easy Steps to a Balanced Science Program* is designed for a specific grade-level range: K–2, 3–8, and 9–12. Each book has the same format and general information, but the specific examples vary according to grade level.

THE *FIVE EASY STEPS*

There are five components to *Five Easy Steps to a Balanced Science Program*, listed here. This five-step structure is a framework that works well with any district-adopted science program, supplemental materials, and assessments currently in place.

Step 1— Establishing an Effective Science Environment

Step 1 emphasizes the management, organization, and strategies for creating an effective and productive science classroom environment that supports student achievement and builds skills and knowledge in students.

Step 2— Problem Solving

Step 2 provides the structure for interactive problem solving, questioning, and higher-order thinking skills within the framework of the standards and objectives. The strategies and activities are correlated with the American Association for the Advancement of Science (AAAS) listing of science process skills.

Step 3—Conceptual Understanding

Step 3 provides the format for teaching the national, state, and local science standards that are essential for student understanding. Students develop content and concept learning around Big Ideas and Essential Questions and the mastery of science information. This becomes the focus for an instructional design that aligns short- and long-range strategic planning and lesson design with end-of-unit and course assessments.

Step 4—Mastery of Science Information

Step 4 presents an organizational tool for weekly science review. There is a focus on literacy and the integration of the best instructional practices for fiction and nonfiction reading, writing, and vocabulary acquisition.

Step 5—Common Formative Assessments

This final step aligns state and school assessments for learning to the science Priority Standards. Common formative assessments are designed, administered, and analyzed within each grade level throughout the year. The data provide predictive value related to student success on district and state assessments. In addition, the school community is involved with understanding the knowledge and skills needed by all students in order to prepare for future learning and involvement in the twenty-first century.

THE NATIONAL ASSESSMENT OF EDUCATIONAL PROGRESS 2005*

The National Assessment of Educational Progress (NAEP) science assessments provide a view of what U.S. students know and can do in science. The assessment was developed by a committee of science and measurement experts to focus on the goals of the NAEP Science Framework. This framework is organized around two major dimensions: the fields of science and the knowing and doing of science. Earth, physical, and life sciences were included in the assessment, which consisted of multiple-choice and constructed-response questions. In addition, hands-on tasks were included.

In 2005, a representative sample of more than 300,000 students in grades 4, 8, and 12 were assessed in science.

- At grade 4, the average science score was higher in 2005 than in earlier years. The percentage of students performing at or above the Basic achievement level increased from 63 percent in 1996 to 68 percent in 2005.

- At grade 8, there was no overall improvement. In 2005, 59 percent of students scored at or above the Basic level.

- At grade 12, the average score declined since 1996. In 2005, 54 percent of students scored at or above the Basic level.

Most states showed no improvement at grades 4 and 8, but five of the 37 participating states did improve between 2000 and 2005. Those states were California, Hawaii, Kentucky, South Carolina, and Virginia.

---

*Source: U.S. Department of Education, Institute of Education Sciences, National Center for Education Statistics. *National Assessment of Educational Progress (NAEP)*, 1996, 2000, and 2005 Science Assessments.

SO WHAT'S STANDING IN OUR WAY?

With the amount of work that is required of a science teacher today, it is difficult to be able to balance the curriculum and standards, instructional methods and assessments, and data collection and review. Many teachers are frustrated yet are committed to providing students with a strong science program but are often overwhelmed with the daily requirements and routines.

Several issues must be addressed in order to meet the challenges of teaching science effectively:

1. Many teachers have not received sufficient and sustainable professional development in teaching science. In some cases, teachers who have no educational background or experience are hired to teach science. These educators need specific professional development, not only in science methods and pedagogy, but also in relationship building, classroom management, and multiple other areas that seasoned teachers take for granted. When given the framework and the tools that build skills and knowledge for teaching science, educators feel more confident in implementing the essential components of a balanced science program.

2. District science programs are confusing, are often not aligned with the state standards or testing, and address too much content to be taught during the course of one year. The science textbook that is chosen by the district many times does not align with the state testing or the class being taught but still becomes the predominant model by which to follow the curriculum. Today's educational arena is full of standards, assessments, and system mandates that govern what and how educators should teach. This can be overwhelming even to the most veteran teacher. There is not enough time in the school year to physically "cover" all of the material in the curriculum and/or textbook. In addition, many teachers perceive gaps in the adopted programs and supplement with additional lessons and activities from other resources. To help educators meet the learning needs of *all* of their students, I suggest that the following structures be developed:

- A framework that incorporates the components of a balanced science program, using the textbook or science series as a resource
- A framework that aligns the standards with best practices in instruction and assessment around a conceptual unit format
- Strategies and activities that support the science process skills, vocabulary, and reading and writing integration

3. Teachers are under enormous pressure to teach science procedures, conceptual understanding, and problem solving AND to prepare students for state tests. Science tends to be a "compartmentalized" subject with multiple units presented to students each year. Although many state curricula tend to spiral or repeat in greater depth at later years, teachers often emphasize what is on the state test and deemphasize the skills and concepts needed for a deep understanding of science. By combining a balance of science understandings and skills, teachers will be implementing the methods that reach all of the components of effective science teaching and learning.

4. Science is often taught in isolation and is not integrated with other subject areas. Teachers who integrate science with other subject areas such as art, music, drama, literature, and history allow their students to go deeper into a scientific understanding of concepts and content. As teachers grow more competent with the science content, the integration of the arts and other core subjects becomes a natural connection for the teacher and the student. The National Science Education Standards support coordinating and connecting science with other subjects, including technology. This integration allows for a more efficient use of instructional time and enhances student achievement and learning.

ORGANIZATION OF THE BOOK

This book is organized into three major parts.

Part One presents each of the five steps, including information on:

- Essential Question(s)
- Description of the step
- Examples of the particular step and its implementation process
- Instructional strategies and activities

Part Two presents grade-level, specific examples with practical strategies for successful implementation inside the classroom.

Part Three provides the reader with time-management ideas and with responses to frequently asked questions about implementing the *Five Easy Steps*.

A section of reproducibles that contain the templates presented in the five steps is provided in the Appendix. These templates may be duplicated for school and classroom use.

The References and Suggested Reading list current resources that the author has found to be supportive of the *Five Easy Steps*, and a Webography lists many current and active Web sites for science.

BALANCING YOUR SCIENCE PROGRAM

As you begin to read through the text and reflect on the current state of your science program, take time to consider the following practices in your school and science classroom.

How are you currently addressing a balanced science program that leads to improved teaching and learning? Take a moment and decide where you are with implementation and performance of the science practices listed in Exhibit I.1.

GETTING STARTED

The implications of the NAEP (2005) study suggest a need for a different approach to teaching science. Book One of *Five Easy Steps to a Balanced Science Program* is for classroom teachers in kindergarten and grades 1–2. It is designed as a framework for implementing the established state or local science curriculum. I encourage every teacher to use the information just as it is presented or adapt it to meet individual teacher and student needs. Hopefully this model will excite and encourage you to create a balanced science program in your classroom. If you have any questions or comments, please do not hesitate to contact me.

Best of luck for a great year! Let's get started with science!

EXHIBIT I.1: Implementation and Performance of Science Practices

Science Practice	Progressing	Proficient	Exemplary	Evidence
My district/school has a balanced science program in place.				
My classroom is a model for effective science instruction, including safety procedures.				
I feel comfortable teaching science.				
My teaching includes the integration of reading and writing in science.				
I read science journals and belong to at least one professional science organization.				
My teaching incorporates the best instructional practices for science labs and experiments.				
My classroom instruction integrates problem solving on a weekly basis.				
My team designs conceptual science units around the Priority Standards and objectives.				
My team uses data to drive instruction.				
My team designs and administers pre- and post-assessments (e.g., common formative assessments)				

"Learn avidly. Question repeatedly what you have learned. Analyze it carefully. Then put what you have learned into practice intelligently."

—EDWARD COCKER, EDUCATOR; AUTHOR OF *ARITHMETICK* (1631–1676)

Reflecting on My Learning !

- What implications does the information in the Introduction have for you as a science educator?

- What are the strengths, weaknesses, and challenges in your school and district?

PART ONE
The Five Easy Steps

STEP 1: Establishing an Effective Science Environment

STEP 1 EMPHASIZES the relationship-building, safety, and management strategies for creating an effective and productive science classroom environment. The teaching and learning should be designed so that students build scientific skills and knowledge with increased student achievement.

Essential Question

How do we design and implement effective science classrooms?

✎ KEY POINTS

1. Effective science classrooms emphasize management, organization, and relationship-building strategies to support student achievement and science-literate students.

2. All teachers should be teachers of science safety rules and procedures.

3. The science lab experience provides students with an opportunity to explore, investigate, and solve problems.

"The astronomer Carl Sagan once said,
'Everybody starts out as a scientist.
Every child has the scientist's sense of wonder and awe.'
Sustaining this sense of wonder presents teachers, parents,
and others close to children with a tremendous responsibility—
and an extraordinary opportunity."

— EVERY CHILD A SCIENTIST: ACHIEVING SCIENTIFIC LITERACY FOR ALL (1998)

DEFINITION OF SCIENCE

Students have their own definition of science. Depending on the grade level, it can range from "cool and awesome" to "gross and smelly." The word "science" comes from the Latin *scientia*, meaning knowledge. According to *Webster's New Collegiate Dictionary* (2009), the definition of science is "knowledge attained through study or practice," or "knowledge covering general truths of the operation of general laws, esp. as obtained and tested through scientific method [and] concerned with the physical world."

At the beginning of the year, teachers need to have an understanding of their students' knowledge of science and specifically of their attitude toward science. Helping students become aware of the work they will be doing in science during the year increases their chance for improved achievement and success. Many teachers have students create their own definition of science during the opening of school activities. These may be displayed around the room or in the hall and will set the tone for a positive beginning to the year.

STUDENT PERCEPTIONS OF SCIENTISTS

No matter how long you teach science, it will become apparent that some students come to class with a high level of knowledge in science and others with limited understanding. The perception of a "scientist" seldom changes over grade-level spans of students. Even if students have had a very skilled teacher the previous year who emphasized the role of science and the scientist, students tend to retain the mental representation of a stereotypical image.

A study done by Mead and Métraux in 1957 found the following about student perceptions of a scientist. The problem is that it really hasn't changed much over the past years:

He is elderly or middle aged and wears glasses . . .
He may wear a beard, may be . . . unkempt. . . .
He is surrounded by equipment . . . and spends
his days doing experiments (p. 387).

At the beginning of one school year, I asked my seventh- and ninth-grade students to draw a scientist. The only thing that basically changed in the drawings was the specificity of the details. Their drawings showed that scientists:

- Wear white lab coats and work in labs that are in basements

- Are usually old, white men with beards and unkempt hair (the Einstein look)

- Smell and like to hold slimy things or test tubes

- Like to play with technology, do experiments, and blow up stuff

- Are really, really smart and use highly sophisticated words

Dealing with Student Perceptions at the Beginning of the Year

The beginning of the year is a good time to introduce students to what science is and what scientists really do. I started with an activity called Build a Scientist. The students brainstorm all of the characteristics of a scientist and these are recorded on the board. Although many of their comments may reflect incorrect assumptions, I encourage them to think about a scientist in today's world and their role in society. These are also included in their list of characteristics. Working in cooperative groups, students use supplies (e.g., cut out a paper shape of a person and use construction paper, glue, string, etc.) to create a visual display of a scientist. Posting these around the room is a way to show students that you value their work, and it makes a great introduction to the beginning of the science year.

THE EFFECTIVE SCIENCE CLASSROOM

The science classroom should be a place of wonder and curiosity. Effective science instructors are continually refining and reviewing the quality of their teaching and of the learning experience for their students. Through the years, I have found that there are three components that must be present for science instruction to be productive for all students. These three components are communication, connection, and consistency.

Communication

Effective science teachers communicate their expectations for management, learning, and achievement on the first day of school. This involves modeling and practicing the procedures that will make the class run smoothly throughout the year. General rules and procedures, including a rationale for each, make students aware of how to behave in all settings. Science teachers have the added responsibility of teaching and practicing safety procedures.

Connection

Some teachers have a gift for connecting with students. They begin on the first day sharing stories and personal experiences with science. Students are able to see that their teacher is actually a teacher and a scientist. Many teachers bring in photos or design slide shows of places they have been that introduce students to their world of science. Teachers who make connections to student learning and personalities establish great relationships with students. Every teacher's goal should be to create a classroom where students work cooperatively, feel safe, and learn at a high level.

Consistency

Effective science teachers are prepared and consistently enforce procedures and classroom management expectations. Student learning and achievement are directly linked to clear and concise planning that is put into effect on the first day of school. Effective science teachers have materials ready, give directions clearly to all students, and have consistent expectations for behavior and learning.

Effective Science Teachers

Do you remember walking into a science classroom as a child? What did you see, hear, and smell? Many of us can remember the class pet, the bubbling sounds of an aquarium, or the smell of dissected material. A classroom that does science should look like a science classroom. And the teacher must have a passion for and love of science in order to transfer this enthusiasm and excitement for the subject to the students.

Effective teachers know how to manage and run a classroom. They know what to teach (the standards and content) and how to teach (instructional strategies), and they have the background knowledge (depth of science understanding) that provides for a powerful learning experience for all students.

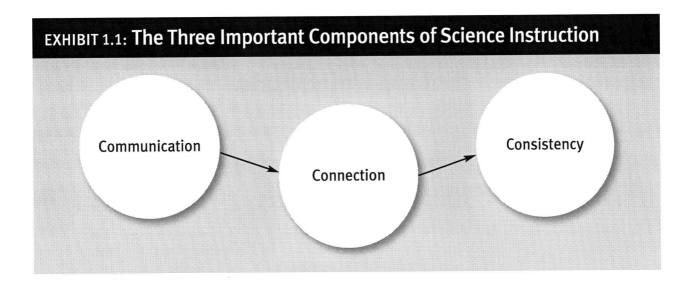

EXHIBIT 1.1: The Three Important Components of Science Instruction

More important is that really effective science teachers go beyond the traditional training in science to acquire a deeper understanding of the concepts and knowledge needed in science. This is accomplished through quality professional development and continual research into the best science literature.

For example, a student who does not understand the cause of the seasons is likely to continue with misconceptions unless the teacher clearly delineates the explanation. This is a content-specific example that has been misunderstood, not only by students but by many adults. The teacher must be able to explain, model, or illustrate such terms as rotation, revolution, and axis in order to have students internalize the concept of seasons. If teachers do not fully understand the seasonal process themselves, there will be a continuation of wrong knowledge.

Effective science classrooms do not just happen; they are created and designed with attention to detail and consideration of all students. There is not enough space in this book to include everything you need to know about establishing a science classroom, but the following three factors must be in place before effective science teaching can occur: building a positive student-teacher relationship, maintaining safety, and managing the classroom (see Exhibit 1.2).

BUILDING A RELATIONSHIP WITH SCIENCE

Classroom Atmosphere

Teaching is about building relationships. Every day I asked myself, "Would I want to be a student in my classroom?" Reflecting on how I responded to my students gave me the opportunity to reevaluate the physical and emotional

EXHIBIT 1.2: Factors Needed for Effective Science Teaching

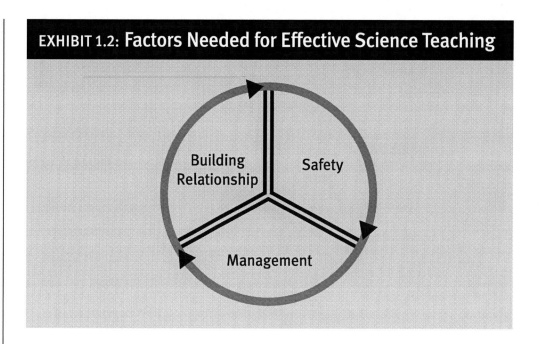

atmosphere in my room. The moment they walk into a classroom, students know what kind of teacher you are. They want to see a clean, colorful classroom filled with items that catch their interest. Posters, lamps, books, stuffed animals, plants, and motivational banners should invite students to enter and be ready to work. It should be evident from the first glance of a classroom that every child is valued and appreciated. A teacher who creates a student-centered, adult-driven classroom has tremendous potential to improve student achievement.

Think back to when you were in school. What did the teacher do to make you feel welcomed, comfortable, and successful? Some teachers have students who can't wait to get to their room, and others have students who look at school as confinement and torture. The most important thing we can do is to create an atmosphere where students are valued, can learn and achieve, and are able to enjoy science.

Classroom atmosphere includes all the things that surround the students: sights, sounds, and smells. Creating an atmosphere that is attractive, functional, and instructionally appropriate is critical to making students feel successful in their work. Student work, bulletin boards, and displays should reflect and help create an atmosphere in which taking risks is acceptable and encouraged and where mutual respect is the daily norm. Teachers must provide a clear definition of high expectations and model behavior to meet those expectations from the first day.

Building Positive Relationships with Students

We know that student success is related to how we treat students in the classroom. Teachers who build positive relationships with their students foster success in student achievement, self-esteem, and lifelong learning skills.

When students leave our classroom after having a chance to participate in science activities, we want all students to:

1. Understand their role as "scientists" in today's world

2. Look forward to learning more about the science around them

3. Express their science learning through a variety of communication methods

4. Practice the science process skills in real-world problem solving

5. Have confidence in their own ability to be involved with science and scientific thought

Teachers set the expectations for student success in science. The teacher's interaction with students and the classroom atmosphere are critical in establishing a positive science learning environment. Students are always curious as to "who" the teacher really is. Many students do not see their teacher as a person. Teachers are perceived as mysterious; they sleep at school and have no life outside of the classroom. Students want to know something that will make the teacher more personable to them. At the beginning of the year, spend time sharing your adventures with science and allowing students to share their own experiences. The following are suggested activities that you can use in the first week of school to create a positive classroom environment that supports student achievement and helps build relationships. The activities are structured so that teachers can connect quickly with students by selecting activities that they feel are most appropriate to their students' grade level and needs.

Meet Your Science Teacher

This activity allows students to get to know you as a teacher. Create a series of multiple-choice questions about yourself. Ask each question and have students discuss possible answers. Give the correct answer and explain why. Sample questions could include: What is my favorite animal? Where is my favorite place to explore? What is my favorite science topic?

Who's in the Science Lab?

This is a fun, interactive activity that students do within the classroom. A series of statements or questions is provided, and time is allocated for the students to mingle among their classmates to find the answers. The names of students are added to the template until all boxes are filled. Time for sharing is encouraged as students learn interesting facts about their classmates. A sample Who's in the Science Lab is found in the Appendix (Reproducibles), in Exhibit A.1.

Science Scavenger Hunt

This activity is designed for students to explore the science classroom and learn the location of important materials. The science process skills are the main focus for this kinesthetic activity as students observe, communicate, measure, and infer about the questions presented. Group students into pairs and give each pair a scavenger hunt worksheet. Allow time for students to investigate the classroom and record their answers. Award a prize to the top groups, if desired. A sample Science Scavenger Hunt is found in the Appendix (Reproducibles), in Exhibit A.2.

The ABCs of Science

The content of science has a preponderance of vocabulary terms and phrases. This beginning-of-school activity is done by partners or teams of students. The ABC template is given to the students and a time limit is established. The objective is to list (and the text may be used) as many science terms beginning with each of the letters of the alphabet. Time to share the answers concludes this activity. A sample ABCs of Science template is found in the Appendix (Reproducibles), in Exhibit A.3.

For an extension of this, students could create visual displays in the form of a graphic organizer highlighting their vocabulary.

Equipment Bingo

All science classrooms use some type of science equipment. Whether they are simple household items or complex chemical apparatuses, students should know the names of the pieces of equipment, how to handle them, and the dangers involved with their use. A template is given to the students with a list of the equipment that will be used during the year. Students randomly write the names in the boxes and the game proceeds like Bingo. The teacher may either call out the name of the piece of equipment or hold up a representative sample. A sample Equipment Bingo template is found in the Appendix (Reproducibles), in Exhibit A.4.

Suggested words, equipment pieces, or visuals for kindergarten and grades 1 and 2 include:

goggles	hand lens	magnet	balance
tripod magnifier	aquarium	microscope	measuring cup
beaker	map	scissors	thermometer
ruler	rain gauge	prism	stopwatch
hot plate	pencil	lab tray	computer

Video Tic-Tac-Toe

With the advancement in technology and media services, students have access to multiple audiovisual materials. Teachers often use video clips to emphasize key points of a topic or concept. All students should be held accountable for pre-, during-, and post-viewing of any media resource. The Video Tic-Tac-Toe is a graphic organizer that is given to students prior to watching a selection. Students are able to share information with a partner, fill in the facts, and then discuss the information. A sample Video Tic-Tac-Toe template is found in the Appendix (rReproducibles), in Exhibit A.5.

BEING A RESPONSIBLE AND SAFE SCIENCE TEACHER

Another component in establishing an effective science classroom is to implement and monitor safety rules and procedures. Science should be a hands-on experience at all grade levels, but science activities can have embedded hazards that come with the work. All equipment, materials, and living organisms must be handled properly and with great respect. Safety in the science classroom is the number one priority of *all* teachers when conducting experiments and demonstrations or when students participate in labs. To ensure a safe science environment, rules, procedures, and specific instructions must be followed at all times.

Teachers should be familiar with the tools and substances that will be used with experiments, demonstrations, and student labs. Although the complexity of the supplies and equipment varies with grade level, there is a need to take responsibility for anticipating any potential danger or hazard. Kindergarten through second-grade teachers should make sure that the use of even the simplest of supplies (e.g., glue, scissors, rulers) is fully understood by the students and that proper rules for handling each object are practiced and monitored. Students in high school should be provided with explicit rules for handling chemicals, following emergency procedures, and displaying proper lab etiquette.

A classic example is from my own teaching experience. I had done an air pressure demonstration in my eighth-grade class. I was very familiar with the tools and how to use them and with the science behind the demonstration. However, there are some things that cannot be controlled and are beyond your power. One of my students decided to "re-create" the experiment at home, caught her bedroom drapes on fire, and burned down her mobile home. Although I had said repeatedly not to do this at home, she was excited about the experiment and wanted to try it for others at home.

What Do You Know about Safety in the Science Classroom?

Safe science classrooms don't just happen. They are well planned, constantly rehearsed, and revised. There is nothing more important than maintaining rules and procedures for science activities. Accidents in a science classroom often occur because students fail to follow the instructions or the teacher is not demonstrating accountability for student safety. If the students are not following the directions, there should be a concerted effort on the teacher's part
to review, reinforce, and monitor student participation in science labs and activities.

One of the best resources for science safety information is Flinn Scientific and its Web site, www.Flinnsci.com/Sections/Safety. This is a complete resource for science materials, supplies, chemicals, and chemical storage, plus free information. Sample safety contracts, chemical data sheets, and a number of free resources are available for download.

Teacher lesson plans should reflect the consistent implementation of a safe science environment. From the start of school to the last day, there should be written evidence that rules, procedures, safety contracts, and behavior expectations are modeled and practiced by both the teacher and all students. Every time you talk about a safety issue, it should be recorded in the lesson plan. A list of the rules and procedures for a specific lab or activity should be recorded on the day that the activity occurs. This supports documentation in case questions arise concerning teacher accountability.

The Physical Environment

All science classrooms will have similar specifications as to appropriate and effective design. Elementary classrooms typically have "science stations" around the room where materials and supplies are accessible to students. Middle schools and high schools often have a dedicated science laboratory or a combination classroom and lab. One of the most important room design components is a minimum of two door exits. The location and appropriate use of emergency equipment such as eyewash stations, safety showers, first-aid kits, fire blankets, and fire extinguishers must be known by the teacher and by all students. All classrooms must have adequate ventilation, along with procedures for dealing with any substance, when doing any science lab experiments. These procedures and other appropriate specifications are required for health and personal safety reasons.

Guidelines, Rules, and Procedures

All science classrooms should have specific rules and procedures for working with equipment, lab stations, and emergency situations. The first week of school should be spent familiarizing students with the nonnegotiable procedures that will be followed during the year.

Materials and Tools

Science is a hands-on subject and can require an enormous amount of materials, equipment, and tools for experiments and labs. Many districts around the country provide elementary school teachers with science kits that are delivered from a central location. In other districts, materials are purchased by individual grade levels based on the current science standards. In order to be an effective science teacher, you must have the necessary supplies to do the hands-on activities. At the beginning of the year, teachers often spend time looking at the resources and correlating the needs to the standards and units prior to ordering supplies and materials.

Visual Displays

All classroom visuals and displays should complement and support the curriculum standards. Students are very visual. Posters and bulletin boards are key in helping students visualize and understand the need for safety procedures in the science lab and classroom. During the first week of school, have students design their own for display in the classroom. Based on teacher topics (e.g., handling equipment, lab work, use of substances), students can create powerful visual aids for posting in the classroom.

Vocabulary terms that are new to students may also be visually displayed around the room. For primary grades, these vocabulary terms may be displayed as visuals so that students can understand the implications of safety procedures.

Along with the visual displays are student-created safety presentations. Using technology, students can design and share PowerPoint slide shows or movies depicting safe science rules and procedures. This empowers the student to take ownership of the safety process and may have more meaning than when it comes from you.

The Science Safety Contract

The student safety contract, which must be signed by both the student and a parent or guardian, is the foundation for safe science classrooms. Grade-level-appropriate rules and procedures are included in the nonnegotiable list of student expectations. No student should be allowed to participate in any science activity until the contract is signed by both a parent or guardian and the student. A sample contract is found in Exhibit 1.3. If you have students who are one to two grade levels above average, you may want to use the sample contract found in Exhibit 1.4.

Safety Quiz

All students should be required to take a teacher-designed safety quiz. The safety quiz is based on the components in the safety contract and on grade-level-appropriate activities. It should include questions on safety procedures, chemical and equipment use, safety materials, emergency procedures, proper

EXHIBIT 1.3: Sample Science Safety Contract: Primary

SCIENCE IS A HANDS-ON EXPERIENCE. There are potential dangers in the lab, and safety is an important part of any scientific work. To help keep our classroom lab safe, the following rules must be followed at all times. No student will be allowed to participate in any science activities until this contract has been signed by both the student and a parent or guardian.

Safety Rules

1. I will listen carefully.

2. I will follow all directions given by the teacher.

3. I will wash my hands after science activities.

4. I will keep myself and others safe.

5. I will properly use and be careful with all equipment.

6. I will be a responsible scientist.

_____   _____
STUDENT SIGNATURE   DATE

_____   _____
PARENT/GUARDIAN SIGNATURE   DATE

_____   _____
TEACHER SIGNATURE   DATE

In case of accident or emergency, contact:

Name: _____   Phone Number: _____

Does your child wear contact lenses?   ❏ yes ❏ no

Is your child color blind?   ❏ yes ❏ no

Does your child have any allergies?   ❏ yes ❏ no

Does your child have any existing medical conditions?   ❏ yes ❏ no

Please list:

_____

_____

EXHIBIT 1.4: Sample Science Safety Contract: Upper Elementary

SCIENCE IS A HANDS-ON EXPERIENCE. There are potential dangers in the lab and safety is an important part of any scientific work. To help keep our classroom lab safe, the following rules must be followed at all times. No student will be allowed to participate in any science activities until this contract has been signed by both the student and a parent or guardian.

Science Safety Agreement

I,_____, WILL:

1. Follow all written and verbal instructions given by the teacher

2. Ask questions, if needed, before beginning a lab procedure

3. Behave in a responsible manner at all times, including no horseplay, joking, or misusing any materials or equipment

4. Use protective equipment for my eyes, face, hands, body, and clothing during all lab experiences

5. Know the location and use of first-aid and emergency equipment

6. Never eat, drink, chew gum, or taste anything in the science classroom

7. Keep my work area clean and free of unnecessary materials, paper, or supplies

8. Clean all work areas and equipment at the end of the experiment. Return all materials and equipment clean and in good working order to the proper storage area

9. Immediately report any unsafe activity by other classmates to the teacher

10. Use proper lab procedures when handling chemicals and substances

11. Handle all living organisms with care and respect

12. Always carry a microscope with BOTH hands

13. Never enter storage areas or chemical closets without permission from the teacher

14. Never remove anything (e.g., chemicals, equipment, supplies, or living things) from the classroom

15. Dress properly (long hair will be tied back; no dangling jewelry or loose, baggy clothing will be allowed)

I,_____, HAVE READ AND UNDERSTAND EACH OF THE STATEMENTS ABOVE AND AGREE TO FOLLOW THEM TO ENSURE MY OWN SAFETY AND THE SAFETY OF OTHERS IN THE CLASSROOM. I AGREE TO FOLLOW THE GENERAL RULES OF APPROPRIATE BEHAVIOR AND TO PROVIDE A SAFE LEARNING ENVIRONMENT FOR EVERYONE.

_____
STUDENT SIGNATURE

_____
DATE

_____
PARENT/GUARDIAN SIGNATURE

_____
DATE

In case of accident or emergency, contact:

Name: _____   Phone Number: _____

Does your child wear contact lenses?   ❑ yes ❑ no

Is your child color blind?   ❑ yes ❑ no

Does your child have any allergies?   ❑ yes ❑ no

Does your child have any existing medical conditions?   ❑ yes ❑ no

Please list: _____

clothing, and behavior. No students should be allowed to participate in any science activity until they have 100 percent success on the safety quiz.

Based on the level of their students, teachers should provide a grade-level safety quiz prior to students doing science investigations. For kindergarten, this may be a set of pictures depicting correct and incorrect procedures; students would circle the correct response. For upper primary students, there may be short phrases, with or without visuals, that represent safe science procedures. In some cases, teachers may actually demonstrate the correct versus incorrect procedures.

Safe, Ethical, and Humane Treatment of Living Organisms

Many classrooms have a pet or living organism that supports science inquiry and investigations. Students must be taught how to care for the animals and plants and why procedures are important to ensure their quality of life. It is advisable to check with your district or school guidelines for having living organisms in the classroom. Some animals are not allowed in some schools. More and more high school biology labs are moving toward virtual dissection through Internet Web sites. Safe preservation of animals and plants is also an important consideration for teachers and students.

Planning for Safe Storage, Handling, and Disposal of Chemicals and Substances

Students will be asked to handle chemicals and substances during science experiments and labs. Primary teachers must be very explicit about how students use even the simplest of materials. Glue, water, food coloring, and paints are often part of science centers, and careful use of these materials will ensure a safe science environment. There are specific guidelines and Material Safety Data Sheets (MSDS) for the twenty most used chemicals, and they are available to all teachers. These can be found on the Internet or in chemical supply catalogs.

The location of chemical storage must meet specific codes, and teachers should be aware of these regulations prior to handling and storing chemicals and substances.

Local school districts will have a plan for disposal of dangerous or hazardous materials. Teachers are required to know this information in case of an emergency, spill, or finding of inappropriate materials.

Parent and Student Responsibilities

Students are required to know and follow the science safety rules at all times. Parents must read and sign the science safety agreement to demonstrate that they understand the need for their child to follow the safety rules and procedures as outlined in the safety contract.

Both parents and students must acknowledge that if one of the parties refuses to sign the safety agreement or the student repeatedly refuses to follow the safety rules and procedures, the student will not be allowed to participate in science labs.

Designing a Safe Science Environment Template

It is important to have a template to follow regarding your plan for safety in the science classroom. Exhibit 1.5 will help you organize your safety system.

THE SCIENCE LAB EXPERIENCE

The science lab can be one of the most powerful experiences that a student can have during class. It is an opportunity for students to practice what scientists do in the real world. In order for labs to be effective, students need to understand not only how to do the experiment but the reason and purpose behind the activity. The lab experience allows students to develop core skills in science, understand problem solving, and improve their attitude toward science and technology.

The National Research Council (NRC) (1998) found that most students are not exposed to effective labs. The NRC study cited a number of reasons for the shortcomings in labs:

- Inadequate or poor school facilities and organization
- Weak teacher preparation
- Poor design in integration of content with process skills
- Cluttered state standards
- Little representation on state tests
- Scarce evidence of what works

Planning for a Science Lab

The most important thing you can do in designing a lab experience is to be well prepared. Basic planning for your lab includes the following:

EXHIBIT 1.5: Template for Designing a Safe Science Environment

Topics	What I Am Doing in My Classsroom
The physical environment (e.g., lab stations, work areas, centers, reading corners)	
Guidelines, rules, and procedures for working safely in the science classroom (e.g., use of chemicals, equipment, behavior, materials) Materials and tools (e.g., microscopes, chemicals, supplies)	
Visual displays (e.g., posters, bulletin boards) depicting safety in science	
Safety contract (e.g., all rules, procedures, consequences, use of equipment and materials, emergency contact information)	
Safety quiz or test (e.g., all rules, procedures, consequences, use of equipment and materials, emergency contact information)	
Safe, ethical, and humane treatment of living organisms (e.g., displays, experimentation, collections, pets)	
Plan for safe storage, handling, and disposal of chemicals and substances (e.g., specific guide-lines, MSDS sheets for twenty most used chemicals, emergency procedures, behavior around chemicals)	

- **Understand what the students must know and be able to do.** This is the teacher's understanding of the standards and objectives. All states and most districts have curriculum standards that outline the core content skills and concepts that students must master before the end of the year. Students must understand the goal of the science activity.

- **Practice and perform all experiments in advance of the students.** All experiments and demonstrations must be fully understood by the teacher. This includes knowing the materials and their use or interaction and the safety issues involved in the lab or demonstration.

- **Understand the science behind the experiment.** Teachers must understand the science that is involved with the experiment or demonstration at the appropriate grade level. Being able to explain the science at the students' grade-level understanding makes learning more relevant to them.

- **Gather and collect all materials for the lab (e.g., equipment and materials, emergency supplies).** This is just common sense and one of the first components of effective lesson design. Having materials and equipment in baskets or tubs is an effective way to organize the supplies needed for a lab or experiment. Secondary teachers often do not have time between classes to "resupply" the materials. Being proactive in planning will ensure a great lab experience for all.

- **Design the student investigation sheet.** All students should have a format for recording the processes and procedures that were used during an experiment or lab activity. The next section, "Four Suggestions for Creating Effective Science Labs," shows several lab format data sheets.

- **Decide how best to introduce lab work to the students.** The teacher should begin the lab experience by modeling proper rules and procedures for completing the lab. This is best done through a demonstration by the teacher, who shares his or her expectations for completing a lab activity. When demonstrating a lab for the students, consider the following:

  - Safety

  - Curriculum standards and related concepts

  - Correct vocabulary and terminology

  - Accurate explanations regarding the outcome

  - Nonverbal and verbal cues

- Assessment and reflection
- Liability

- **Identify the related science process skills.** Students must be required to go deeper into problem solving and thinking at their grade level. There should be a direct correlation between the purpose of the activity and the science process skills used in completing it.

- **Integrate the science.** Science should not be taught in isolation but presented so that students make connections with prior knowledge. Conceptual units designed with curriculum connections to other content are powerful learning tools for students. Consider areas such as art, music, drama, and history when presenting the science standards during the year.

- **Share discussions and self-reflections.** All lab experiences should include a time for shared discussion, questions, and self-reflection from the students. Verbal and nonverbal communication is an essential component at the end of a lab activity. Various strategies, such as journaling and cooperative group sharing, give students a chance to review and revise their thoughts.

Four Suggestions for Creating Effective Science Labs

The following ideas from the National Research Council (2005) will help you create an effective science lab:

- **Have a clearly stated purpose.** Students must understand the goal of the lab in order to effectively conduct the experiment.

- **Use an integrated approach.** The science content must not be presented in isolation; it must be connected to what the students already know and what they must be able to do.

- **Emphasize the science process skills.** Students must be asked to go deeper into problem solving and thinking.

- **Use discussion and self-reflection.** Students must have the opportunity to communicate their thoughts, both verbally and in writing.

Students love to do lab experiments and activities. Effective science labs take time and practice. The teacher must plan the lab based on the appropriate grade level and needs of the students. Exhibit 1.6 provides a suggested template for designing a science lab.

EXHIBIT 1.6: A Suggested Teacher Planning Template for Designing a Science Lab

Lab Title:

Description	
Standards **Objectives**	
Text Resources	
Essential Question	
Big Idea	
Science Process Skills	
Materials/Equipment	
Time	
Procedure/Set up	
Curiosity Hook	
Student Investigation Sheets	
Assessment	

THE SCIENCE LAB REPORT

Lab reports are an essential component of an effective lab and are designed to help students with organizing the lab activity. All students should be able to complete a lab report based on an appropriate grade-level activity; in this case, kindergarten through grade 2.

Student lab reports generally have the following in common:

- **Title and title page.** Not all lab reports require a title page, but all have a title for the activity. Students should include the title of the experiment, their name and date, and the course name. The title should be concise and reflect the main point of the experiment; for example, "Effects of Sunlight on Green Plants," or a title that is appropriate to the grade level.

- **Statement of the problem.** This includes the question that is being asked or the purpose of the lab. In some lab reports, this may be an introductory paragraph explaining the purpose or objective of the lab. It may contain a brief amount of background information and why the lab was conducted.

- **Hypothesis.** The student must make an "educated" guess as to a possible solution to the problem. There is a direct relationship between the variables and the predicted results that is based on what is being studied.

- **Materials.** All labs require specific materials and equipment. This is an inclusive list that must be prepared prior to beginning the lab.

- **Procedures.** Procedures are listed in clear steps, and each step is numbered and written in complete sentences. The procedure must be clear enough so that someone else could replicate the lab. Steps are outlined sequentially and contain detailed explanations, appropriate to the grade level.

- **Data.** Any tables, charts, observations, and notes that were made during the lab are included in this section. The data are professional looking with appropriate titles and labels. If calculations are used, they are shown correctly and labeled appropriately.

- **Results and/or conclusion.** Students will either accept or reject their hypothesis. Many lab reports have a single paragraph that summarizes what happened in the experiment.

Exhibit 1.7 shows a sample template for a lab report for primary grades. If you have students who are one or two grade levels above average, you may use the template for upper elementary grades in Exhibit 1.8.

EXHIBIT 1.7: Sample Lab Report for Primary Grades

Name: **Date:**

Problem:

I think . . .

I saw . . .

I know . . .

EXHIBIT 1.8: Sample Lab Report for Upper Elementary Grades

Name: **Date:**

Title:
Purpose/Problem:
Introduction:
Hypothesis:
Materials/Equipment:
Procedure:
Observations and Data:
Conclusions/Summary:
Further Ideas:
References:

THE ROLE OF CLASSROOM MANAGEMENT

Classroom management is the single most difficult challenge for a beginning teacher. When it is combined with a teacher's lack of science skills and content knowledge, the situation presents an ongoing need for effective professional development. Classroom management refers to everything a teacher does to organize time, space, and students so that effective instruction occurs every day. It was, however, described by one of my first-year science teachers as "herding mosquitoes." Teachers who learn to manage the classroom keep students on task and actively engaged with their work. It takes time to develop the skills for managing student behavior, and conversations and demonstrations in this area should be both an individual and team effort.

Being well prepared for class will reduce the amount of time spent on behavioral issues. What happens on the first day of school determines the pace of success during the rest of the year.

Robert Marzano in *What Works in Schools* (2003, pp. 88–89) defines classroom management as "the confluence of teacher actions in four distinct areas: (1) establishing and enforcing rules and procedures, (2) carrying out disciplinary actions, (3) maintaining effective teacher and student relationships, and (4) maintaining an appropriate mental set for management. Only when effective practices in these four areas are employed and working in concert is a classroom effectively managed."

According to Walker, et al. (1996), "Classroom systems are developed by teachers to support the larger school-wide policies and procedures and to manage the academic performance and social behavior of students within instructional environments and arrangements" (p. 198).

Posting the rules, rewards, and consequences provides students with a reminder of the teacher's expectations. It is important to establish a working set of procedures from the beginning so that students understand the way the class will operate. Schools and grade-level teams often develop and implement a consistent behavior management plan that is structured for classroom operation.

Managing a Science Classroom: Three Simple Strategies for Establishing Science Procedures

There is no shortage of books, resources, and materials on classroom management. In the science classroom, I have used and encouraged other science teachers to use the following three strategies at the beginning of the year. Each deals with establishing an effective science environment, setting behavioral expectations, and providing teacher-directed science instruction. Again, the number one expectation for all science activities is safety for everyone in the classroom.

Communication, connection, and consistency (discussed earlier in this chapter) are accomplished with these classroom-management strategies, which all grade-level teachers can use and that encompass visual, auditory, and kinesthetic learning.

The Focus Board

The front board of the classroom should be designed to contain information related to the lesson of the day. Signs or placards, consistently used on a daily basis, are placed on the front board to alert students to the day's activities (see Exhibit 1.9). These should be located in the same place every day so that students become accustomed to seeing them and following through with expectations for their work. Such signs could include:

- The date
- The standard/objective (the "unwrapped" version)
- Essential Question(s)

EXHIBIT 1.9: Example of a Focus Board

Today's Date
Focus Lesson
Warm-Up
Objective
Homework

- The warm-up

- Homework

The Spot

This strategy is used when giving directions, providing teacher-directed information, and beginning or ending class. The teacher stands in a particular location in the front of the room where all students can see him or her. This location is where all students focus their attention while the teacher provides directions, instructions, and information. If the teacher moves off of "the spot," then students know that it is time to work cooperatively or independently.

Six-Step Directions

The six-step direction format is used to guide students as they begin their work. Teachers should use the "See It, Say It, and Write It" model when they give directions to students in order to accommodate multiple learning styles. Use these starter words when giving oral directions:

- WHEN . . .   "In one minute . . ."
- WHO . . .   "Everyone in the class . . ."
- WHAT . . .   "Will move to their lab stations . . ."
- ASK . . .   "What questions do you have . . . ?"
- ACTION . . .   "You may quietly move and get started . . ."
- TIME . . .   "You have twenty minutes."

DIFFERENTIATION IN THE SCIENCE CLASSROOM

Teaching in a regular classroom has become more challenging than ever today. Differentiated instruction is not a single strategy but a process approach to instruction that matches student characteristics to teacher instruction and assessment. It is more important than ever to implement science instruction in ways that accommodate the variety of academically diverse learners in our classrooms. Students are expected to achieve at high levels and with a standards-based curriculum; therefore, we must find ways to ensure that all students have equal opportunities for success.

Science teaching lends itself to strategies that support differentiated in-struction. It is based on best practices in education and incorporates science inquiry, critical thinking, problem solving, and different learning modalities. The learning needs of the students drive the instructional planning of the teacher. Science teaching involves recognizing the fact that individual students and cooperative groups of students can use different content, processes, and products to achieve the same understanding of science concepts.

Differentiating provides multiple opportunities for students to acquire information, use scientific process skills, and demonstrate understanding of what they have learned.

Choices in Science Differentiation

Science teachers, with effective planning, can provide multiple opportunities for students to learn information, understand science concepts, and demonstrate what they have learned. Skills and concepts can be differentiated in several ways, as shown in Exhibit 1.10.

What Does Differentiation Look Like in the Science Classroom?

Science instruction and effective differentiation require clear, focused objectives. These are linked to the science Priority Standards, and when teachers differentiate, they do so in response to a student's readiness, interest, and/or learning style. Readiness refers to the skill level and background knowledge that the child brings to the class. Interest refers to topics that the student may want to explore or for which he or she may want to build additional scientific content. A learning style includes the preferred modalities for learning (e.g., visual, auditory, linguistic, or kinesthetic), grouping (e.g., individual, small group), and environmental factors (e.g., space, loud, quiet).

Differentiation Strategies

Exhibit 1.11 offers a variety of strategies that teachers may use to differentiate science instruction in their classroom.

THINGS TO CONSIDER

Science is a hands-on subject. Establishing an effective science classroom environment requires attention to detail. Routines and practice allow students to understand and participate in a structured atmosphere with high expectations for success. ∎

EXHIBIT 1.10: Differentiation in Science Skills and Concepts

Instructional Science Choice:	Differentiation According to:
Content:	Student readiness, interest, learning style
Process:	Student readiness, interest, learning style
Product:	Student readiness, interest, learning style
Environment:	Learning preferences (auditory, visual, kinesthetic, linguistic) Class participation (individual, small group, whole group)
Assessment:	Written (common and summative assessments) Verbal presentation Multimedia performance tasks Problem-solving tasks

EXHIBIT 1.11: Strategies for Differentiating Science Instruction

Strategy	Definition	Example
Flexible Grouping	Students should be part of many groups during the year based on their interest, academic ability, and learning modality. These groups may be heterogeneous or homogenous, but both allow for collaboration and cooperation skills. The teacher may allow students to self-select their group, or he or she may assign them.	Different levels of text, based on student reading and comprehension ability, allow for exploration of science topics based on student need. Small-group labs may be implemented on a variety of content, process, and product levels.
Interest/Learning Centers	Centers are used to provide enrichment opportunities for students who demonstrate mastery with the required work in class. The center activities can be differentiated by the level of content, cognitive thinking skills, and product design.	An elementary teacher may have a center designed on science-related children's literature. Activities, aligned with the books, could provide students the chance to explore and go deeper into a specific topic or content.
Tiered Assignments	Tiered assignments, based on various levels of activities, are designed to ensure that students investigate ideas at a level that builds on their prior knowledge and that promotes additional learning. The curriculum standards and objective(s) are the same, but the level of process and product complexity is varied according to the student's level of readiness. Tiered assignments make great learning or interest-center activities.	Bloom's taxonomy is used to create content, process, and product activities based on science topics. For example, a unit on the ecosystem would include a variety of resource materials at differing levels and related to different learning modalities. The task involved can be adjusted by complexity, challenge levels, and modification for special-needs students.

EXHIBIT 1.11: Strategies for Differentiating Science Instruction

Strategy	Definition	Example
Questioning	All students should be accountable for information and higher-order thinking. Students need to be challenged with more than basic questions. Wait time is one of the most important questioning skills that allows for students to process information before being required to answer the question.	There are several strategies for questioning in Chapter 3 that teachers can use to effectively implement questioning in their classroom.
Compacting	Compacting is the process of adjusting instruction based on a student's prior mastery of the learning standard(s) and objective(s). Compacting consists of three steps: (1) student assessment to determine the student's level of knowledge of the material to be studied and what needs to be mastered; (2) implementation of a plan for what the student needs to know, and excusing the student from what he or she already knows; and (3) time for enrichment and acceleration.	A student who already knows the principles of aerodynamics would be allowed to develop and conduct additional experiments related to the topic, while other students are given more direct instruction on the concept. Compacting is not a common strategy in elementary school because most students are just being exposed to the nature of science.
Choice Grids	Choice grids are organizers that provide a variety of activities that students choose to complete during a specified amount of time. Teachers can organize the grid so that students focus on targeted skills or concepts during this work time.	A Tic-Tac-Toe grid is filled with activities based on simple to complex activities that address a variety of content, process, and product tasks. Students must complete a set of activities based on their location in the grid.

"I LOVE science! It's my favorite subject at school, and my strongest.
Science is a great example of life and people.
It's also a great learning experience."

– B.J., STUDENT

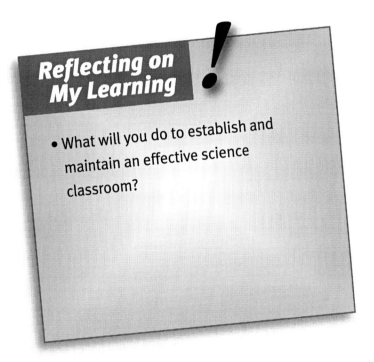

Reflecting on My Learning

- What will you do to establish and maintain an effective science classroom?

STEP 2: Problem Solving

CHAPTER 2

S TEP 2 PROVIDES THE STRUCTURE for interactive problem solving, questioning, and higher-order thinking skills within the framework of the state and local standards and objectives. The strategies and activities are correlated with the American Association for the Advancement of Science's (AAAS) listing of science process skills.

Essential Question

How do we provide opportunities for students to apply and explain scientific knowledge?

✎ KEY POINTS

1. Problem solving allows students to understand the process of scientific thinking.

2. Scientific reasoning, problem solving, and the science process skills are the keys to gaining effective scientific knowledge through real-world challenges.

WHAT IS PROBLEM SOLVING?

Problem solving is the core of scientific exploration and investigation. This provides students with the opportunity to apply their science skills and knowledge. Effective problem solving requires that students are presented with or identify a problem, follow a series of steps to a solution, and communicate their findings to others. During the experimentation, students incorporate the science process skills within the steps.

The problem-solving step of the balanced science program addresses two areas:

1. The application of science in the Conceptual Understanding Unit to the standards and process skills

2. The communication of scientific thinking to others

Problem solving is the ability to find a solution to an uncertain or difficult situation. Learning to solve problems requires practice and focused reading and thinking skills. Students are either given a real or simulated problem and must use critical-thinking skills to solve it. They will need to have an integrated approach and draw from a variety of disciplines. Problems that have personal relevance to the students are excellent choices because they encourage active learning, participation, and retention of science information.

Problem: A question raised for inquiry, consideration, or solution
Solve: To find a solution, explanation, or answer for
(*Merriam-Webster Dictionary Online*, 2010)

SCIENCE AS INQUIRY

Children, especially in the early grades, are naturally curious about the world around them. Inquiry-based instruction provides the opportunity for students to engage in scientific investigation, improve critical and creative thinking skills, and understand the role that science and scientists play in today's society.

Science as inquiry "involves asking a simple question, completing an investigation, answering the questions, and presenting the results to others" (National Research Council, 1996, p. 122). Inquiry refers to those processes and skills used by scientists when they make investigations.

Until about 1900, science education was regarded as teaching students a series or collection of facts through memorization. Science, however, should be taught as a way of thinking about and doing science. One of the ways to effectively do this is to have students actively participate in science through labs and hands-on experiences. Providing students with experiential activities allows them to explore science content and pose questions.

Rationale for Science as Inquiry

The National Science Education Standards (1996) recommend that students:

• Experience the richness and excitement of knowing about and understanding the natural world

• Use appropriate scientific processes and principles in making personal decisions

• Engage intelligently in public discourse and debate about matters of scientific and technological concern

• Increase their economic productivity through the use of the knowledge, understanding, and skills of the scientifically literate person in their careers

Five Essential Features of Science Inquiry

In an article from the National Research Council (2000b) entitled *National Science Education Standards: A Guide for Teaching and Learning,* a list of five essential features that support science inquiry is presented. These are:

1. Learners are engaged by scientifically oriented questions.

2. Learners give priority to evidence, which allows them to develop and evaluate explanations that address scientifically oriented questions.

3. Learners formulate explanations from evidence to address scientifically oriented questions.

4. Learners evaluate their explanations in light of alternative explanations, particularly those reflecting scientific understanding.

5. Learners communicate and justify their proposed explanations.

The teacher's role in designing a classroom where questions are frequently asked before, during, and after science investigations is critical for improving student thinking and knowledge development. However, it is important to remember that good teaching involves multiple methods and techniques and that there is no one way to teach. Science cannot always be considered a "step-by-step" process but should be a subject where multiple avenues of exploration and investigation are used to solve problems.

Primary students have an intense curiosity about science and the natural world. Teachers of these students must have the ability to quickly differentiate their instruction based on student interest and need. Teachable moments happen at all grade levels. My definition of this is "anything that absolutely, positively disrupts the instructional flow of the lesson but makes a significant impact on learning." In a kindergarten class, the students had been exploring a unit on animals and habitats. During the science "show and tell," one student explained to the class that he had brought something to share. He proudly stood up and pulled a worm out of his pocket. He said, "I have lots more in my pocket. You want to see?" When the teacher told him that the worms would all die and that this wasn't a good home for the worms, he replied, "They are okay. I put dirt in there for them." The concept of animal care and habitats was totally understood by the students. However, the teacher took advantage of this teachable moment to revisit the topic of animal needs. The worms were put in the safety of a jar, along with the dirt, and were added to the science observation center.

THE SCIENCE PROCESS SKILLS*

The basic science process skills are what we do when we conduct scientific exploration and experimentation. The science process skills form the foundation for scientific methods. There are six basic science process skills:

1. Observation
2. Communication
3. Classification
4. Measurement
5. Inference
6. Prediction

As teachers select the Priority Standards and objectives, they should determine which of the process skills to include with their teaching. Research indicates that when science process skills are a specific and planned outcome of a science program, students learn those skills. Studies focusing on the Science Curriculum Improvement Study (SCIS) and Science—A Process Approach

---

* Source: American Association for the Advancement of Science, 1989.

EXHIBIT 2.1: Science Skills

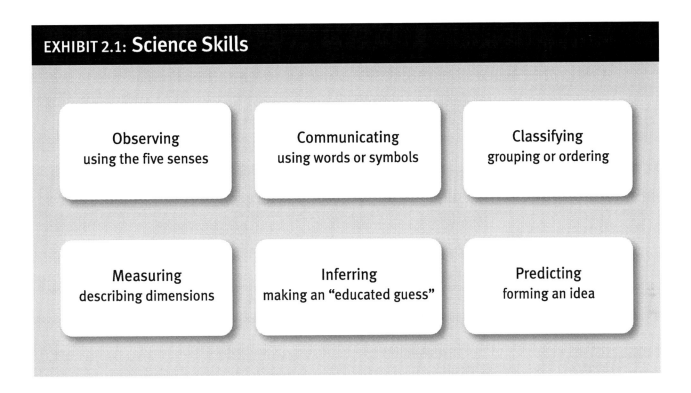

Observing
using the five senses

Communicating
using words or symbols

Classifying
grouping or ordering

Measuring
describing dimensions

Inferring
making an "educated guess"

Predicting
forming an idea

(SAPA) show that students, if taught process skills abilities, not only learn to use those processes but retain them for future years in education. There must be a deliberate incorporation of the process skills in the science classroom.

> *"If you can't explain something simply, you don't understand it well. Most of the fundamental ideas of science are essentially simple, and may, as a rule, be expressed in a language comprehensible to everyone. Everything should be as simple as it can be, yet no simpler."*
>
> — ALBERT EINSTEIN, THEORETICAL PHYSICIST (1879–1975)

Problem Solving and the Science Process Skills

The scientific method and scientific thinking have described science skills for years. The AAAS developed these skills in the 1960s. Today, they are accepted throughout science education as being the building blocks of inquiry and investigation. The science skills are shown in Exhibit 2.1.

SELECTING THE PROBLEM-SOLVING TASK

There is a wealth of information in science textbooks and on the Internet for selecting appropriate problem-solving tasks. The References and Webography sections at the end of the book provide additional resources.

Problem solving in science depends on the student's ability to work through the process of scientifically solving a multistep problem. When the teacher aligns the problem-solving task to the instructional standards in the conceptual unit, students have the chance to demonstrate their ability to apply their scientific learning to a real-world problem.

Teachers who select a problem for students to solve should look for one that allows students to demonstrate their ability to apply science to a real-world situation. Consider the following questions when selecting the problem:

1. Does the problem promote application of the science concepts presented in the current instructional focus or unit of study? **Does it relate to current unit focus?**

2. Does the problem correlate with the students' instructional level? **Is it understandable to all students?**

3. Is the problem relevant and engaging to the students? **Does it capture students' interest?**

4. Does the problem require higher-order thinking processes? **Does it challenge scientific thinking?**

5. Is the problem integrated with other content areas? **Does it involve ideas from more than one science standard?**

6. Do I, as teacher, fully understand the concepts, vocabulary, and methodology that are involved in this problem? **Do I understand all of the science involved?**

The Teacher's Role

The teacher has a critical role in making students successful in becoming independent problem solvers. As the first step in problem solving, the teacher must emphasize scientific reasoning and the use of scientific terminology and vocabulary in the problem. Careful modeling of the steps shows students how best to address a problem. This process takes time, and teachers should be ready to provide assistance and support as needed.

Exhibit 2.2 should be used as a teacher planning tool for developing a problem-solving activity. It is very important that the teacher understand the science behind the problem so that scientifically sound explanations can be given to students. A problem-solving example for elementary grades is provided in Exhibit 2.3, and Exhibit 2.4 lists a number of possible problem-solving tasks.

THE INSTRUCTIONAL SEQUENCE

Once the problem-solving task has been determined, the teacher should follow a specific instructional sequence to teach students how to scientifically solve the problem and how to communicate the procedure and the results. It is essential that students experience success with their first problem-solving tasks if they are to develop independence with problems at a later time. Teachers must deliberately provide opportunities for students to solve problems and communicate their results. Three approaches should be used:

- Whole class

- Independent work

- Cooperative teams or pairs

Whole-Class and Independent Work

The first step of the instructional sequence involves the teacher modeling the problem for the whole class. The students must understand how the problem is connected to the current unit or topic of study. By presenting a variety of problem-solving methods, the teacher guides the students through a series of steps to solve the problem.

1. The teacher and students read the problem together. The problem is discussed, and the teacher makes sure that students fully understand the problem and understand what kind of solution is required.

2. The teacher presents several strategies for problem solving. Students may select a strategy based on their learning style or modality. Teachers who use problem solving on a regular basis in their classroom find that students approach each problem with a variety of personal methods. In order to become independent problem solvers, children must practice until they identify one or two specific strategies that work for them and that transfer to real-life problem-solving situations. When students are working through problems, suggest to them that they:

- Use manipulatives

- Act it out

- Find a pattern

- Guess and check

- Simplify

EXHIBIT 2.2: Template for Planning a Problem-Solving Task

Problem-Solving Title:

Standard(s):	
Concepts:	
Skills:	
Scientific Vocabulary:	
Science Process Skills:	
Problem/Essential Question:	

EXHIBIT 2.2: Template for Planning a Problem-Solving Task

Problem-Solving Title:

Evidence of Problem:	
Materials/Equipment:	
Procedures:	
The Science Behind:	
Assessment:	

EXHIBIT 2.3: Problem-Solving: Elementary Example

Problem-Solving Title: Sink or Float

Standard(s):	• Sort objects by observable properties, such as size, shape, color, temperature (hot or cold), weight (heavy or light), texture, and whether objects sink or float.
"Unwrapped" Concepts:	• Observable properties (size, shape, color, temperature, weight, texture, sink, float)
"Unwrapped" Skills:	• Sort
Scientific Vocabulary:	Matter, physical property, size, shape, color, temperature, weight, texture, sink, float
Science Process Skills:	Observing, measuring, predicting, inferring, communicating, classifying
Problem/Essential Question:	Why do some objects float and others sink?

EXHIBIT 2.3: Problem-Solving: Elementary Example

Problem-Solving Title: Sink or Float

Evidence of Problem:	Student will be able to determine the characteristics in an experiment that cause objects to float or sink.
Materials/Equipment:	Goggles Clear carbonated beverages Spaghetti Macaroni Raisins Dried corn Dried beans
Procedures:	Collect the materials needed for the demonstration. Using the problem-solving data sheet and the problem-solving task write-up page, follow the directions. Do the demonstration using the above materials with the carbonated beverage. The student will do the experiment with raisins.
The Science Behind:	The carbon dioxide bubbles will adhere to the surface of the raisins (it is rough). The bubbles increase the volume of the raisin substantially, but not the mass. The overall density of the raisin is lowered, causing it to be carried upward. Archimedes' Principle states that the buoyant force exerted on a fluid is equal to the weight of fluid being displaced. Because the raisins have a greater volume, they displace more water. The buoyant force of the surrounding fluid is what pushes the raisins to the top of the container. When the raisins reach the top, the bubbles pop when exposed to air. The raisins become more dense, and they sink. The cycle repeats as more bubbles stick to the raisins; they become less dense and are pushed back up.
Assessment:	This is the student problem-solving task write-up sheet based on a problem-solving rubric.

EXHIBIT 2.4: Sample Problem-Solving Tasks

Topic	Activity	Sample Variables
Physics	Balloon Rockets	The type of balloon The size of the balloon The length of string The type of string The amount of air in the balloon
Physical Science	Static Electricity	The type of balloon The temperature of the water
Earth and Space Science	Cloud Watching and Weather Predicting	The time of day The type of cloud The daytime temperature
Biology	The Wilting Garden Plants	The soil quality The amount of water The humidity The rate of transpiration The nutrients The type of plant
Chemistry	Tarnished Pennies Cleaned with Taco Sauce	Age of the penny Type of sauce Brand of sauce Type of ingredients

- Use an illustration
- Make a list

3. Students work five to ten minutes to try to solve the problem using one or more of the problem-solving strategies.

4. Students record their individual work on paper. A data sheet, provided by the teacher, helps guide the student's thinking and problem solving.

5. Students share possible solutions with a partner and then with the teacher.

6. The teacher records the answers for the class to see.

7. Students record the answers and decide on the best solution to the problem.

Cooperative Teams or Pairs

Once students are familiar with the problem-solving process, the teacher assigns students to small, cooperative teams of three to four. However, this time the teacher presents the problem, ensures understanding, and allows the teams to move forward with the solution. The students again use a data sheet to record their work. Members of the group are allowed to share their solutions with the rest of the class. A discussion about the different solutions determines the actual answer.

Modeling and sharing are critical components of effective problem solving. Students may choose to display their answers around the room and use them as references for additional activities.

"Equipped with his five senses,
man explores the universe around him
and calls the adventure science."

— Edwin Powel Hubble, 1954

SUGGESTED STEPS FOR TEACHING PROBLEM SOLVING

To make the teaching of problem solving easier for teachers, a sequence of steps is suggested to help students communicate orally and in writing about the process used in solving the problem.

1. Introduce the problem to the class and make connections to the topic or unit of study. Begin the problem with a couple of starter statements or questions, such as "Today we are going to . . ." or "How does _____ relate to _____?"

2. Read the problem together. Read the problem out loud with the students or have them follow along on their paper.

3. Create questions related to the problem. Student interest is generated with some warm-up questions, such as "What do you think will happen?" or "What is this?"

4. Tell students to research the problem. Students research the problem by collecting information, facts, and data about the problem. The teacher should understand the science behind the problem. If for some reason you don't know the answer to a student's question, then reply that you don't know, but that you and the class can work on finding the answer together.

5. Have students construct a hypothesis. This is an opportunity for students to create their own question about the problem and try to create a hypothesis about the answer. You may ask guiding questions, such as "If I do _____, what do you think will happen?"

6. Allow students time to attempt to solve the problem individually, record their work, and share possible solutions. If the problem is presented as a demonstration, have students respond by either writing or drawing a possible solution. If the problem is presented as a lab or paper activity, students will use their data sheet to record their responses.

7. Make a decision about the solution. The teacher must decide on a possible solution; however, keep in mind that students may find alternative answers that are also valid. The solution should reflect the content or topic that is being learned through the conceptual unit.

8. Students complete a data or task sheet. Students begin their problem-solving task by recording initial solutions on a data sheet. (A template for a Problem-Solving Task Data Sheet can be found in Exhibit A.6 in the Appendix.) Once answers are shared with the class, the student completes a problem-solving task write-up.

9. Data sheets are shared with the class. After students have completed their data sheets, have them share their work with the class or their cooperative group.

PROBLEM-SOLVING STEPS FOR PRIMARY GRADES

Teachers must model the suggested problem-solving steps (listed in the previous section) at the beginning of the year so that all students have a solid understanding of the process. Teachers must consider the age-appropriate and developmentally ready level of their individual students and adjust these steps as needed. It is helpful to post the list in the classroom or provide a copy for students to place in their science journal.

Problem-Solving Task Write-Up Guide

One of the most important components in problem solving is the ability to communicate the work in writing. The Problem-Solving Task Write-Up Guide (Exhibit 2.5) provides a format through which students can organize and display the work that was done on the data sheet. Exhibit 2.6 is a template for the write-up, while Exhibit 2.7 shows an example of a problem-solving activity used in first grade. The problem is presented along with a sample write-up.

TAKING TIME FOR PROBLEM SOLVING

Teachers should take time for students to understand the problem-solving sequence presented here. Many of the problem-solving activities done by teachers can be modeled through a demonstration and performed by pairs or small groups of students. Teachers should spend time showing students how to correctly complete their data sheet and the problem-solving write-up. This will ensure that the process is clearly understood by all students.

The problem-solving step is designed for students to communicate their science understanding using a structured format and specific procedure. The ultimate goal of problem solving is to have students who can transfer the process and procedures into real-world science.

CREATIVE PROBLEM-SOLVING CHALLENGES

Wesley was one of my favorite students. He was a quiet, respectful African-American student in my highly gifted earth science class. Wesley had lots of friends and was always involved in group work and projects for the class, but he never spoke out or volunteered to talk aloud. I realized that I wasn't challenging Wesley in class, but I knew that he loved to build models and work on things, so one day I suggested that he bring in something that he had made. When he did, we were amazed at what we saw! I'm not sure what it was, but it

EXHIBIT 2.5: Problem-Solving Task Write-Up Guide: Primary Grades

Directions:

1. Write your name and date on a piece of paper.
2. Solve the problem using words, pictures, or a combination or both.
3. Number each step as you work to solve the problem.
4. Write an explanation to match your problem.
5. Explain what science process skills were used in solving the problem.
6. Write your answer in a sentence under your solution.
7. Write a short paragraph that explains how you solved the problem (not for kindergarten or first grade).
8. Make sure that you include evidence of science vocabulary.

EXHIBIT 2.6: My Problem-Solving Task Write-Up: Primary Template

Name: **Date:**   **Title of Problem:**

Important Science Vocabulary:

Hypothesis (What I Think):

Steps to the Solution:

What Actually Happened:

EXHIBIT 2.6: My Problem-Solving Task Write-Up: Primary Template

Name: **Date:**   **Title of Problem:**

My Understanding:

Solving the Problem

1. Use your favorite strategy to solve the problem.

2. Number or label your steps.

3. Write a sentence or two telling how you solved the problem.

4. Go back and check to make sure that you are finished.

Problem-Solving Write-Up

Use the problem-solving write-up template to record your information. Make sure that you include science vocabulary in your report.

EXHIBIT 2.7: Student Work Sample for Problem Solving

Problem-Solving Task Write-Up: Primary

Title of Problem: Dancing Raisins

Sink or Float

Materials:

- Goggles
- One can of Sprite
- One box of raisins

Will the raisins sink or float? Why?

Important Science Vocabulary:

sink, float, carbon dioxide, density

What I Think:

I think that the raisins will sink to the bottom of the glass.

My Steps to Solving the Problem:

First, I put on my goggles and got the materials.

Then, I opened the box of raisins but didn't eat any.

Next, I poured the Sprite into the glass. I had to be careful not to let it fizz up too much.

Then, I dropped in 6 raisins and watched them. I watched the raisins sink at first, and then bubbles attached to the raisins.

EXHIBIT 2.7: Student Work Sample for Problem Solving

Problem-Solving Task Write-Up: Primary

What Actually Happened?

The raisins started floating up and down in the Sprite.

My Answer:

Raisins don't sink in Sprite. Bubbles help them float. The bubbles are made of carbon dioxide and stick to the raisins. This makes them float.

The Science Process Skills I Used:

When I did this experiment, I was using scientific thinking. I used predicting when I thought about what might happen with the raisins. I used observing as I watched the raisins. I inferred when I made a guess. I communicated my answer in writing.

My Understanding:

I understand that some things sink and some things float. Some do both. It is cool!

But I wonder about:

What would happen if I used peanuts or macaroni instead of raisins? Would they do the same thing? What if I used diet Coke or Pepsi? Would it matter?

made noise, blew smoke, moved, and flashed lights. The questions from his classmates poured in, and Wesley proudly took time to answer each one—and we still wanted more. After several weeks, one of the other students asked Wesley how he got to be so smart. Wesley said, "I read," and nothing more. I wish I could have had a video of this moment, because it symbolized everything that I would want for all science students.

We learned more from the quiet nature of a self-motivated child than from all of the content in the 679 pages of the earth science textbook. Wesley became a high school chemistry teacher and did well until he moved on into another career. The curiosity, the enthusiasm to explore and experiment, and the constant quest for knowledge are what we want to have our students learn in science class. My most memorable Wesley invention was a blackboard that the teacher could actually write on when giving notes or information to the class. I only wish that Wesley had patented the "Smartboard" back in the 1980s.

"Creative problem solving is a process of creating a solution to a problem. It is a special form of problem solving in which the solution is independently created rather than learned with assistance."

"Creative problem solving always involves creativity. However, creativity often does not involve creative problem solving, especially in fields such as music, poetry, and art. Creativity requires newness or novelty as a characteristic of what is created, but creativity does not necessarily imply that what is created has value or is appreciated by other people."

"To qualify as creative problem solving, the solution must either have value, clearly solve the stated problem, or be appreciated by someone for whom the situation improves."

—Wikipedia, en.wikipedia.org/wiki/Creative_problem_solving (July 19, 2009)

"The most beautiful thing we can experience is the mysterious. It is the source of all true art and science."

—Albert Einstein, theoretical physicist (1879–1975)

For years after teaching Wesley, I tried to provide time for my students to experience creative challenges in class.

For those of you who want to try the Creative Challenge Activities, several are listed in the Appendix in Exhibit A.7. I did these in my classroom once a month, often on a Friday after a great week of work.

As with all science experiments and labs, safety and appropriate grade-level activities are the most important aspects of these problems. Whether you are a kindergarten teacher or a high school chemistry teacher, make the challenges fit your curriculum and needs.

A list of materials, the problem statement, and the solution are given with each challenge. Preparation and collection of the materials (per group) ahead of the day's events will ensure a smooth beginning for the activity. Teachers also have the right to change any of the suggested materials depending on their supplies and needs.

Many of these challenges have been adapted from resources involving problem solving and scientific thinking (see the Webography at the end of the book for suggested sites to find challenges).

PROBLEM-SOLVING SCORING GUIDES

A *scoring guide* (also known as a rubric) is an assessment tool used to judge the quality of a student's performance in relation to content-specific standards. Scoring guides provide a consistent set of guidelines to rank student work and to describe a range of possible student responses. Teachers who use scoring guides can provide:

- Feedback about student progress toward meeting an understanding of the standard
- A common vocabulary for discussing the standards across grade levels and across districts
- A vehicle for student self-evaluation and self-reflection
- A strategy for peer feedback among students

Scoring guides enable both the student and teacher to identify a set of performance criteria that indicate the level of proficiency reached on a particular sample of student work. The scoring guide must be clearly articulated with specific language that everyone understands, and it must be created collaboratively with the students.

A scoring guide used in science must describe the performance in specific, measurable, attainable, relevant, and observable terms. It is important that the scoring guide be applicable to problem solving week after week. This avoids the "too many rubrics" over time.

Issues to Address in Creating Problem-Solving Scoring Guides

There are several issues to address in creating problem-solving scoring guides. It is obvious that you want the students to give you the "right" solution to the problem and to follow the problem-solving write-up guidelines. In addition, the student may provide multiple strategies or ways to solve the problem and still provide the correct solution. The inclusion of appropriate science vocabulary should be considered in the problem-solving write-up. It is a teacher's decision whether writing conventions such as spelling, grammar, punctuation, and sentence form will be assessed.

Note: Proficiency must address the *science* first and foremost. Students should have ongoing opportunities to reflect upon and revise their work with feedback from the scoring guide.

Guidelines for Writing the Problem-Solving Scoring Guide

The following guidelines will help you to write an effective scoring guide:

• Make it general but with specific language.

• Refer to write-up guideline requirements.

• Decide how many performance levels to include.

• Start the rubric with a "3" or "proficient."

• Provide a copy for the students *before* they do the problem-solving task.

• Determine what "proficient" performance is for the task.

• Chart out your first draft.

Steps in Designing a Science Problem-Solving Scoring Guide

A predetermined sequence is used to assess a problem-solving task, as follows:

1. Decide on the number of performance levels. It is recommended that four levels be used as the optimum for creating the scoring guide.

2. Select a format that is either holistic or analytic in form. Holistic scoring guides simply list all of the criteria for a particular level. These are easier for the student to use. For the beginning of problem solving in science, it is recommended that the holistic scoring guide be used. Analytic scoring guides represent all of the criteria for each performance level in a chart format. Each of the problem-solving components (i.e., data sheet, scientific process, science vocabulary, task write-up) is represented by specific criteria for each level of the scoring guide.

3. Identify the problem-solving components. The teacher first develops a list of the "nonnegotiables" that must be included on the scoring guide. Once you have identified these items, decide under which performance level they should be placed. It is important that this be done as a grade-level team or a content team of teachers to ensure consistency in scoring across a grade level. Some items to consider are:

- Correct answer

- Wrong answer but appropriate procedure/process

- Scientific reasoning and inclusion of science vocabulary

- Proof or defense of answer/solution

- Following the data sheet and write-up guideline

4. Start with "proficient." Teachers must decide what "proficient" means during problem solving. It is suggested that the criteria be the "nonnegotiables" of the work but not be regarded as minimum or standard. Remember that these must be specific, measurable, attainable, relevant, and observable indicators.

5. Determine what "exemplary" means. This refers to what the student does that is "beyond" the "proficient" level. It may include multiple ways to solve the problem, the use of additional scientific vocabulary, or the use of multiple materials.

6. Determine what "progressing" and "beginning" mean. These levels are also described in relation to the "proficient" level.

Exhibit 2.8 shows a problem-solving scoring guide that may be used in kindergarten to grade two or as a guide for teachers to design their own. Once the scoring guide is finalized, it can be used on every problem-solving task during the year.

Assessing Students

A simple problem-solving scoring guide provides students with specific feedback on their work. Students should be able to contribute to the scoring

EXHIBIT 2.8: Problem-Solving Scoring Guide: Primary

Name: **Title of Problem:**

Exemplary

All "proficient" criteria *plus*:

Written work explains step-by-step process used to solve problems

Proficient

Correct answer

Solves problem on data sheet with words, pictures, and/or numbers

Includes problem statement

Follows all problem-solving guide directions to complete write-up

Progressing

Meets three of the "proficient" criteria

Beginning

Meets fewer than three of the "proficient" criteria

Tasks to be repeated after remediation

Self-Evaluation:

guide development process during class by stating what great work would look like in a problem-solving activity. It is important for teachers to provide additional opportunities for students to reach the "proficient" level. Giving students the scoring guide prior to the problem-solving task enables students to understand what is required for a "proficient" performance on the task.

Teachers can provide feedback related to following directions, using a problem-solving method, showing a visual representation, and indicating the steps to solving the problem on students' data sheets. Students should be able to include a written explanation. Students must use correct science vocabulary in their final write-up that corresponds to the problem.

PROBLEM-SOLVING AND DISCREPANT EVENTS

Every other Monday in my class was D.E. (Discrepant Event) Day. My sole purpose with this plan was to increase science interest in my students, integrate the science process skills and scientific inquiry, and teach the problem-solving method. Although the discrepant event did not always align with the current science standards, the outcome of this planning was excitement, enthusiasm, involvement, and an increase in science skills and knowledge. Looking back, I would now align the discrepant event with the conceptual unit, but years ago there were no standards or established curriculum.

A **discrepant event** is something that surprises, startles, puzzles, or astonishes the observing student and, in some cases, adults. The "basic rules of science" do not appear to be consistent, and the outcomes are unexpected or conflict with what the student predicts before the event. Discrepant events motivate students to investigate the science concept. This strategy can be used in all science classes and should be aligned with the standards and objectives being taught at the time. These teacher demonstrations encourage good problem solving on the part of the student. It is important for the teacher to present the discrepant event as a science problem that the class will be investigating or trying to solve. The true science concept is not immediately revealed to the students but is one that evolves over the teaching of the conceptual unit.

Discrepant events can be used:

• To engage students in inquiry

• As a demonstration followed by discussion to introduce a new science concept

• To engage students in science process skills

• As a warm-up to stimulate curiosity and critical thinking

• As a take-home lab activity (when proper safety procedures are outlined)

• As a challenge for students to create additional or similar investigative labs

The Value of Discrepant Events

Discrepant events are fun and engaging for students. For years after you have taught students, they will return and remember some of your demonstrations. Discrepant events encourage students to:

• Observe and write about their observations in a science journal

• Pay attention to detail in all types of observations

• Formulate questions that could be researched or that could lead to investigations

• Seek out answers to those questions through observation, exploration, and research

• Record and share information learned

Ten Things to Remember If Doing a Discrepant Event

1. Practice, practice, practice, practice. The secret to doing discrepant events in a science class is the amount of practice the teacher has completed before doing the demonstration. Many discrepant events have a special technique for making them successful, and practice is the only way to ensure an accurate performance.

2. Focus students' attention on the topic. All materials should be prepared ahead of time and be visible (depending on the discrepant event) to the students when they enter the room.

3. Pre-assess what students already know about the topic. A discrepant event introduces a new science concept. Activating prior student knowledge before doing the demonstration is a good way to get the students interested in the activity.

4. Inform the students about the objectives and science process skills they will use. The discrepant event is not about teaching science content first. The teacher should focus on the science process skills (e.g., observing, inferring, communicating, and predicting) and not on the actual science standard for this activity.

5. Think about the questions you will ask. Teachers should develop several questions that will help guide the learning process. One of the purposes of doing discrepant events is to arouse curiosity and interest by presenting a problem. These questions would focus on the materials, what the teacher is doing, and what might happen.

6. Conduct the event for the students. Position yourself and the materials so that all students can see the demonstration.

7. Plan no more than ten minutes from start to finish. Some discrepant events will require student helpers to assist the teacher with the demonstration in order to keep the event within a given period of time.

8. Tell students what you are doing during the discrepant event. Active participation is important as students watch a discrepant event. Teachers may start with a guiding question, such as "What do you see?" or "If I do ____, what do you think will happen?" Students may record their answers in their science journals and expand on the activity later.

9. Start the demonstration with a question. Make the discrepant event relevant to the students. This must be age appropriate so that students begin to see connections between this demonstration and examples in the real world.

10. Decide whether or not to explain the answers. This is left up to the teacher. When students complete the conceptual unit, they should be able to explain the discrepant event based on their knowledge of the science concepts. I still have students whom I taught thirty years ago ask me "How you do dat?" in reference to a discrepant event that I had done in class. Even though they had a thorough understanding of the science concept and could explain the science behind the discrepant event, they were fascinated with the logistical steps that made the event work.

During the discrepant event, students will:

• Observe the discrepant event and write about their observations in a science journal

• Pay attention to detail

• Formulate questions that could be researched or that could lead to investigations

• Seek out answers to those questions through observation, exploration, and research

• Record and share the information learned

Exhibit 2.9 is an example of a discrepant event.

EXHIBIT 2.9: A Discrepant Event

Science Process Skills

- Observing
- Inferring
- Predicting
- Communicating
- Measuring

Sample Discrepant Event

This discrepant event may be used for multiple standards in science. It is a K-12 example of a demonstration that can be adapted to grade-level standards involving science inquiry, problem solving, process skills, weather, gravity, and many other science strands.

This discrepant event serves only as an example for teachers who want to implement this strategy.

Title of the Discrepant Event: Floating in Air

Time: 10 minutes

Related Science Standards/Objectives: *Source*: North Carolina Public Schools (www.ncpublicschools.org)
North Carolina Standard Course of Study for Science, Competency Goal 2 (2.01, 2.02)

"Unwrapped" Skills and Concepts:

SKILLS	CONCEPTS
Conduct	Investigations
Use	Appropriate tools
Investigate	How moving air interacts with objects
Describe	How moving air interacts with objects
Observe	Force of air pressure

Materials

One hair dryer   One Ping-Pong ball

EXHIBIT 2.9: A Discrepant Event

Opening Questions

What are some things that float? Why do they float?

Can Ping-Pong balls float? Where can they float?

What characteristics of a Ping-Pong ball will make it float?

Can a Ping-Pong ball float in the air?

How could we make a Ping-Pong ball float?

How can we use the hair dryer to make the Ping-Pong ball float?

Presentation Steps

Plug in the hair dryer and turn it on.

Put it on the highest setting and point it straight up.

Place your Ping-Pong ball above the hair dryer and watch what happens.

Scientific Explanation

The floating Ping-Pong ball is an example of Bernoulli's Principle, which is the same principle that allows heavier-than-air objects like airplanes to fly. Bernoulli, an eighteenth-century Swiss mathematician, discovered that the faster air flows over the surface of something, the less the amount of air that is pushed on the surface. If you position the ball just right, the air flows evenly around each side. The moving air forces the ball upward and gravity pulls the ball downward. This is due to air pressure.

Challenge

Try different sizes of hair dryers and air-filled balls such as a beach ball.

See if you can float two or three Ping-Pong balls at the same time.

For those who are adventurous, use a roll of toilet paper and a yard blower!

Discrepant Event Competition

I decided to do the discrepant events in the form of a contest. I spent time researching and finding appropriate discrepant events that I could perform easily, and I used simple materials that did not require a great deal of cost. Prior to the performance, I gave each student a Discrepant Event Contest Form (see Exhibit 2.10). I did the demonstration and they had until Friday to submit their answers to me. This was a wonderful opportunity for them to conduct some research in the media center, at the library, or on the Internet. On Friday, they would hand in their forms and we would discuss the results.

I am leaving the grading or assessing to individual teacher discretion for this activity. Although it took time to acquire the collection of discrepant events, I found this to be one of my students' favorite days of the week. They often came to school just to see "How she do dat?" in class.

In order to give you more assistance with discrepant events, a planning template is provided in Exhibit 2.11. In Exhibit 2.12, you will find a number of actual discrepant events that will provide fun and learning in your classroom.

EXHIBIT 2.10: Discrepant Event Contest Form

NAME _____

DATE _____

What you observed: _____

_____

_____

The explanation: _____

_____

_____

Source of information: _____

_____

_____

EXHIBIT 2.11: Discrepant Event Planning Template

Title of the Discrepant Event:

Time:

Science Process Skills:

Related Science Standards/Objectives:

Materials:

EXHIBIT 2.12: Discrepant Events for Classroom Fun and Learning

Title: How Can We Make Ice Disappear? (GRADE LEVEL K–2)

Concept:

Weather and the water cycle

Materials:

Heat source (hot plate)

Pan of ice

Procedures:

1. Show students the pan of ice and ask them to describe the ice.

2. Ask them to explain how we could make the ice "disappear."

3. Place the pan on the hot plate and allow the ice to melt.

4. Continue evaporating the ice.

Explanation:

The ice goes through the steps of the water cycle. At first it melts and changes to water. Then the water evaporates into the air. If you are able to collect the evaporating water, you can demonstrate precipitation to the students.

EXHIBIT 2.12: Discrepant Events for Classroom Fun and Learning

Title: The Missing Animal (GRADE LEVEL K–12)

Concept:

Animal adaptation, fossil record, science inquiry

Materials:

Two stuffed dinosaurs (or any two animals)

Procedures:

1. Show the two animals to the class.

2. Have them discuss the characteristics of each and how they are both alike and different.

3. Tell the following story:

One day, two animals were walking down a very, very muddy path and stopped at the edge of a stream. It was so muddy that you could clearly see their footprints. As the animals met, something happened. When the paleontologist discovered the footprints, there was only one set of footprints on the other side of the stream.

What happened?

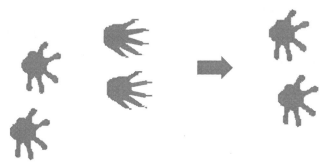

EXHIBIT 2.12: Discrepant Events for Classroom Fun and Learning

Title: The Singing Wine Glass (GRADE LEVEL K–12)

Concept:

Sound, vibration, resonance

Materials:

One long-stemmed wine glass

Vinegar

Water

Procedures:

1. Wash your hands with soap and water and place the wine glass on the table.
2. Dip the tip of your finger into the vinegar, hold the glass by the stem, and rub the rim with the dipped finger. Do this slowly.
3. Pour a little water into the wine glass and repeat step 2.
4. Have students listen and formulate a hypothesis as to the creation of the sound.

Explanation:

This is similar to hitting a glass with a hard object. Once the first few vibrations are produced, they make the glass resonate because the pitch is the same as that of the glass. By adding water, the vibrating mass increases and the pitch is lowered.

EXHIBIT 2.12: Discrepant Events for Classroom Fun and Learning

Title: The Spinning Top of Water (GRADE LEVEL K–12)

Concept:

Simple machines

Materials:

One round toothpick

Eyedropper with water

Circle cut from poster board (four inches diameter)

Procedures:

1. Insert about one-third of the toothpick through the center of the circle to form a top.

2. Fill the eyedropper with water.

3. Ask for a student volunteer (he or she may get a little wet).

4. Have the student spin the top.

5. Hold the eyedropper above the top and drop water onto the spinning disk.

6. The assistant, and maybe you, may get a little wet.

Ask: "Why does my assistant get wet?"

Explanation:

The dropped water is thrown away from the spinning top. If the motion of the water was slowed down, you could see that the water leaves the top in a straight line. This information is used to design a simple machine that will pump water. Water that flows onto a spinning disk creates a pump that can move water from one place to another.

EXHIBIT 2.12: Discrepant Events for Classroom Fun and Learning

Title: Sink or Float (GRADE LEVEL K–12)

Concept:

Density

Materials:

A large, clear container with water

Miscellaneous samples of rocks (one must be pumice)

Procedures:

1. Fill the container with water so that all students can see.

2. Show them the samples of rocks.

3. Ask them if rocks float or sink.

4. Place the samples in the water, one at a time, leaving the pumice for last.

Explanation:

Pumice is an extrusive igneous rock that has a lower density than an equal volume of water. Pumice forms from an airborne, rapidly cooled, frothy magma that is full of tiny gas pockets, making it extremely light in weight. There is enough trapped air in pumice that it is buoyant. Use caution to not soak the pumice or it will fill up with water.

EXHIBIT 2.12: Discrepant Events for Classroom Fun and Learning

Title: Swinging in Time (GRADE LEVEL 3–12)

Concept:

Revolution

Materials:

Several small balls

Miscellaneous lengths of string (e.g., one foot, three feet, six feet)

Strong tape

Procedures:

1. Cut the string to various lengths.

2. Attach a string to each of the balls.

3. Spin the shortest string and ball around your head.

4. Spin the others, one at a time, with increasing length.

5. Count the time it takes the ball to make a complete rotation around your head.

6. Ask students to watch and predict why counting out loud increases with each different ball.

Explanation:

As the location of the planets goes farther from the sun, the time for one revolution increases.

EXHIBIT 2.12: Discrepant Events for Classroom Fun and Learning

Title: Floating in Air (GRADE LEVEL K–12)

Concept:

Bernoulli's Principle

Materials:

One hair dryer	One Ping-Pong ball
One yard leaf blower	One beach ball
One roll of toilet paper	One broom handle

Procedures:

1. Plug in the hair dryer and turn it on.
2. Put it on the highest setting and point it straight up.
3. Place your Ping-Pong ball above the hair dryer and watch what happens.
4. Do the same for the beach ball with the yard blower.
5. Do the same for the toilet paper (put on a broom handle).

Explanation:

The floating Ping-Pong ball is an example of Bernoulli's Principle, which is the same principle that allows heavier-than-air objects like airplanes to fly. Bernoulli, an eighteenth-century Swiss mathematician, discovered that the faster air flows over the surface of something, the less the amount of air that is pushed on the surface. If you position the ball just right, the air flows evenly around each side. The moving air forces the ball upward and gravity pulls the ball downward. This is due to air pressure.

EXHIBIT 2.12: Discrepant Events for Classroom Fun and Learning

Title: The Lighted Bulb (Grade level K–12)

Concept:

Conductivity, circuits

Materials:

A battery (D cell will work) and masking tape

A 1.2-volt bulb and bulb holder

Three short wires (paper clips will work)

Variety of materials (some metal and some nonmetal)

Procedures:

1. Connect one wire of the bulb to the negative pole of the battery using masking tape.

2. Connect the other wire to the other pole of the bulb holder.

3. Tape one end of the third wire to the positive pole of the battery using masking tape.

4. The two loose ends are the test terminals. Test the light, and when the two loose ends are touching, the bulb should light.

5. Test a variety of materials and have students determine which type of material is a good conductor.

Explanation:

This demonstration shows the importance of a completed circuit. When a conductor is placed between the terminals, the circuit is complete and the light goes on. When a nonconductor is placed between the terminals, the electrons cannot flow and the light stays off.

EXHIBIT 2.12: Discrepant Events for Classroom Fun and Learning

Title: Floating Soda Cans (GRADE LEVEL K–12)

Concept:

Density

Materials:

One large container with water

Four to five different cans of soda (diet, regular, cherry)

Procedures:

1. *Prior to students entering classroom:* Cover the soda cans so that the labels are not visible.
2. Fill the container with water. Tell students that you have a variety of sodas and want observations and hypotheses as to the activity.
3. Place the cans in the water and observe the differences.

Explanation:

Diet sodas usually contain aspartame, an artificial sweetener, while regular sodas use sugar. In addition, there is a little bit of space, called "head space," above the fluid in each can of soda. This space is filled with gas, which is much less dense than the soda; thus, it "floats" above the soda. It is this space above the soda that lowers the density of diet drinks just enough to make them float. Sugared drinks still have this head space, but the excessive amounts of sugar added make the can denser than water. Both sugar and aspartame are denser than water. It is a matter of how much of each is used. The 41 grams or so of sugar added to a can of regular soda makes it sink. The tiny amount of aspartame used in diet sodas will have a negligible effect on the mass, enabling the can to float.

EXHIBIT 2.12: Discrepant Events for Classroom Fun and Learning

Title: Balloon Breath (Grade level 3–12)

Concept:

Air pressure

Materials:

Two students

Two two-liter plastic bottles

Two balloons

Ice pick or compass end

Procedures:

1. *Prior to students entering to see this demonstration:* Punch three small holes in the bottom of one of the plastic bottles (students should not be able to see the holes).

2. Give a balloon and a plastic bottle to each student.

3. Ask them to place the balloon inside of the bottle with the top of the balloon attached to the lip of the bottle. Have each student try to blow up the balloon.

4. Ask the class why one student can blow up his or her balloon and the other cannot.

5. Develop a hypothesis for the action.

Explanation:

In the bottle without the holes, the balloon won't inflate much because the bottle is already filled with air. There is no room for the balloon to expand. In the bottle with the holes, the air molecules will exit through the holes, thus allowing the balloon to inflate. You can also cover the holes with your fingers and the balloon will stay inflated.

EXHIBIT 2.12: Discrepant Events for Classroom Fun and Learning

Title: The Mysterious Bubble (GRADE LEVEL K–12)

Concept:

Homeostasis, structure and function of cells, cohesiveness of water molecules

Materials:

Clear, very large test tube

Water

Rubbing alcohol

Rubber stopper

Food coloring (optional)

Procedures:

1. Fill the test tube one-half full of water and add a drop or two of food coloring.

2. Fill the test tube to the top with the alcohol.

3. Stopper the test tube.

4. Invert over and over. A "bubble" will form in the test tube.

Explanation:

The water and ethanol molecules are different sizes, with the ethanol molecules being smaller.

Some of the ethanol fits in the spaces between the water molecules.

EXHIBIT 2.12: Discrepant Events for Classroom Fun and Learning

Title: Magic Pepper (GRADE LEVEL K–12)

Concept:

Surface tension, density, cohesiveness of water molecules

Materials:

Clear bowl with water

Ground black pepper in a shaker can

Whole peppercorns

Eyedropper with dishwashing liquid

Procedures:

1. Place the bowl of water on the overhead.

   Ask the students, "If I drop the peppercorn into the water, what will happen?"

2. Drop the peppercorn.

3. Ask the students, "If I sprinkle some ground pepper in the water, what will happen?"

4. Sprinkle a small amount of ground pepper onto the water.

   Ask the students, "Why didn't this pepper sink?"

5. Ask students to observe the reaction when a drop of soap is added to the center of the pepper.

6. Add one drop to the center of the bowl.

7. Develop an explanation for the action.

Explanation:

The density of the peppercorn is greater than the density of the water (thus it sinks). The ground pepper is lighter and is held up by the surface tension of the water. The water molecules form a film that can also hold up light objects such as a needle, paper clip, and razor blade. When you add a drop of detergent, the cohesive forces between the water molecules are broken, the film is removed, and the pepper sinks.

EXHIBIT 2.12: Discrepant Events for Classroom Fun and Learning

Title: Sunken Ice Cube (GRADE LEVEL K–12)

Concept:

Density, mass, volume

Materials:

Two clear cups

Rubbing alcohol

Water

Two large ice cubes

Procedures:

1. *Prior to students entering the room:* Fill one cup with water and one cup with rubbing alcohol.

2. Set up the cups so that students can make observations about each one.

3. Place one ice cube in each cup and have students record observations and predict the outcome.

Explanation:

The alcohol has a lower density than water. This is why the ice sinks in the alcohol and floats in the water.

It is dependent on the relative density of the object compared to that of the liquid.

EXHIBIT 2.12: Discrepant Events for Classroom Fun and Learning

Title: Skewered Balloon (GRADE LEVEL K–12)

Concept:

Polymers

Materials:

Several helium-quality balloons

One bamboo skewer or knitting needle

Petroleum jelly or cooking oil (optional)

Goggles

Procedures:

1. *Prior to students:* Dip the tip of the skewer in the petroleum jelly.

2. Blow up the balloon.

3. Carefully poke the skewer through the balloon without popping it.

   Make sure you use the top and bottom ends of the balloon.

4. Display the skewered balloon as students enter the room.

5. Ask students how this can happen.

Explanation:

You must find the part of the balloon where the latex molecules are under the least amount of strain. If you could see the rubber that makes up a balloon on a microscopic level, you would see many long strands or chains of molecules. These long strands of molecules are called a polymer, and the elasticity of these polymer chains causes rubber to be stretchy. Blowing up the balloon stretches these strands of polymer chains. You must insert the skewer into the balloon at the top and bottom, not through the middle.

EXHIBIT 2.12: Discrepant Events for Classroom Fun and Learning

Title: The Cartesian Diver (Grade level K–12)

Concept:

Properties of air, density, sinking, and floating

Materials:

One two-liter, clean soda bottle

Water

Eyedropper

Procedures:

1. Fill the soda bottle almost to the top with water.
2. Fill the eyedropper halfway with water and test it in another container of water (it should float just below the waterline).
3. Place the eyedropper in the bottle and seal the bottle with the bottle top.
4. Gently squeeze the bottle at the bottom. If the eyedropper has the right amount of water, it will sink when the bottle is squeezed and float when the bottle is released.

Explanation:

The air inside the eyedropper is compressed; the water level rises and the dropper sinks. By releasing the pressure, the water is pushed out of the eyedropper; it gets lighter and floats again.

EXHIBIT 2.12: Discrepant Events for Classroom Fun and Learning

Title: The Bleeding Paper (GRADE LEVEL K–12)

Concept:

Acids and bases

Materials:

One piece of goldenrod paper

Ammonia-water solution

Paper towels

Procedures:

1. Conduct this experiment in a well-ventilated area away from students but close enough for them to see the demonstration.
2. Mix a mild solution of ammonia and water (household ammonia works well).
3. Place the piece of goldenrod paper on a clean, flat surface.
4. Talk to the students about the importance of using clues to find solutions to problems. Not all evidence is visible.
5. Dip your hand into the ammonia-water solution; shake off the excess.
6. Quickly place your hand, palm down, on the goldenrod paper.
7. Have students observe what happens.

Explanation:

The term *goldenrod* is used to describe a color of paper. Some brands of goldenrod paper contain a special dye that turns bright red in solutions that are basic (like the ammonia-water solution). The paper will turn back to yellow with an acidic solution (like vinegar or lemon juice).

EXHIBIT 2.12: Discrepant Events for Classroom Fun and Learning

Title: Crushing Cans (GRADE LEVEL K–12)

Concept:

Gas laws

Materials:

Several clean soda cans

Hot plate or Bunsen burner

Cold water in a bowl

One tablespoon of water

Tongs

Procedures:

1. Fill the soda can with about one tablespoon of water.

2. Place the can on the hot plate and heat it until you see steam.

3. Quickly use the tongs to invert the open can into the bowl of cold water.

Explanation:

The can was filled with water and air before you heated it. By boiling, the water changed states from a liquid to a gas, which is called water vapor. The water vapor pushed the air out of the can. When the can was turned upside down and placed in the water, the water vapor condensed and turned quickly back to water. This small amount of water cannot exert much pressure on the inside of the can, so the pressure of the air pushing from outside is great enough to crush it.

EXHIBIT 2.12: Discrepant Events for Classroom Fun and Learning

Title: Disappearing Water (GRADE LEVEL K–12)

Concept:

Conservation, absorption

Materials:

Three Styrofoam cups

Water

Sodium polyacrylate (also found as Aqua Slush or Supergel)

Procedures:

1. Place about one teaspoon of sodium polyacrylate in one of the cups.

2. Now create the illusion.

3. Pour water into one of the other cups.

4. Move it around (just like the hidden peanut trick at the circus) and have the students guess where the water is.

5. Pour this water into another cup (not the one with the Supergel).

6. Move it around and again ask the students to guess where the water is.

7. Pour the water into the last cup (the one with the Supergel).

8. Move it around and ask students to guess again.

9. When you invert all of the cups, the water will not flow out.

10. Don't hold the cup upside down too long as the gel will fall out.

Explanation:

This powder instantly turns a liquid into a solid. Sodium polyacrylate absorbs from 800 to 1,000 times its weight in water and is the secret ingredient that's used to absorb "liquid" in baby diapers.

SCIENCE FAIRS

Science fairs provide students with the opportunity to apply the scientific method and process skills to conduct independent research. The results are presented in a school-wide or locally sponsored event where the student's efforts are displayed and where students are interviewed to determine scientific knowledge. Many times students are able to advance to state, national, and international competitions.

As recommended by Carl Tant (1992), the most universally used format includes:

1. Title page
2. Abstract
3. Introduction
4. Literature review
5. Experimental procedures

6. Results
7. Discussion of results
8. Acknowledgment of assistance
9. References

Five Easy Steps does not address the science fair process in any depth but encourages teachers to read and explore information about how to implement an effective science fair in their classroom or school.

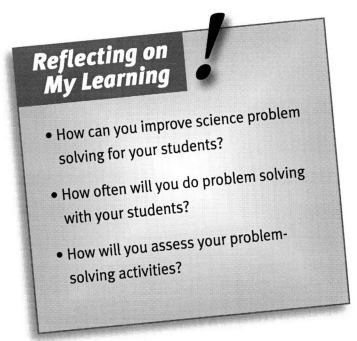

Reflecting on My Learning

- How can you improve science problem solving for your students?

- How often will you do problem solving with your students?

- How will you assess your problem-solving activities?

STEP 3: Conceptual Understanding

Designing a Conceptual Unit

S TEP 3 IDENTIFIES the national, state, and local science standards that are essential for student understanding. Students develop content and concept learning around Big Ideas and Essential Questions and the mastery of science information. This becomes the focus for an instructional design that aligns short- and long-range strategic planning and lesson design with end-of-unit and course assessments.

Essential Questions

What is conceptual understanding?
Why is it important in science?

✎ KEY POINTS

1. A conceptual approach to learning science allows students to deepen scientific understanding by connecting science concepts to science meaning.

2. Engaging students in best instructional practices for collaboration and reading scientific information provides students with the opportunity to expand their skills and knowledge in science.

A BASIC UNDERSTANDING OF SCIENCE INSTRUCTION

There is a basic understanding among scientists and science education experts that science instruction should promote meaningful understanding of science concepts, processes, and thoughts. Students should not just be memorizing scientific facts or vocabulary but developing a global understanding of how science works. The state standards present teachers with a long litany of goals, objectives, and benchmarks that do not support conceptual understanding. This is why science is so "departmentalized" and promotes the unit-to-unit teaching. With a more concise, structured approach to science teaching, students would be more productive in their learning environments and with science opportunities.

Dr. Dorothy Gabel is a professor of science at Indiana University. In an article entitled "Conceptual Understanding of Science" (2003, p. 70), she states:

> If we want to improve the conceptual understanding of science, teachers must be selective in the concepts they include in instruction. Much depends on the background that students bring to the course. This means not that more content should be moved to lower grade levels but that the National Science Education Standards (1996) should be used as a guide to provide reasonable levels of content. The Standards carefully delineate what leading experts in the field deem appropriate for most students at each grade level. Increasing the content may force students to memorize and turn out to be detrimental to conceptual understanding.

RATIONALE FOR THE USE OF CONCEPTUAL UNITS

A conceptual unit in science is an effective process for educators to move beyond the conventional practice of using the science textbook as a guide on what to teach and how to access student learning. Students do not retain information if they do not have the chance to participate in meaningful instruction and learning in science.

Developing conceptual understanding allows the teacher to provide meaning-based instruction based on Priority Standards around a central topic. Because of the complex nature of science vocabulary, teachers should strive to make connections between units so that students internalize the content.

Step 3 begins with the identification, with district or state science standards, of a particular grade-level topic that is essential for student understanding. That topic becomes the focus of the conceptual science unit that is deliberately aligned with an end-of-unit assessment.

THE VALUE OF COLLABORATIVELY PLANNING A CONCEPTUAL UNIT

Teams of teachers across the country are faced with an overwhelming amount of information about effective teaching and student learning. Many schools are classified "schools in crisis" by national, state, and local criteria. I have had the opportunity to work with many of these schools over the years. It is interesting to have watched the transition from individual, compartmentalized instruction to Performance Learning Communities and Data Team formats. One of my favorite schools was on every national and state watch list. Student achievement was dismal in all grades and stagnant in the gifted students, and teacher morale was at an all-time low. The school decided to implement several components that would ensure consistency and continuity across the grade-level and content areas. Because teaching is an isolated profession, the idea of teaming was very new to this school, and obstacles to overcoming the "my room, my lesson" mindset had to be addressed in the beginning. From the first day of implementation, the school was committed to making a difference for the students. The hard lesson was that the teachers had to "look inward" and address personal issues before moving forward. The school implemented a conceptual unit design process to move beyond the traditional textbook and more into identifying Priority Standards, best instructional practices, and common formative assessments.

CONCEPTUAL UNIT DESIGN

Science has typically been a subject in which students memorize isolated scientific facts with little understanding of broad, interrelated concepts. The new standards movement is asking students to demonstrate a deep understanding of a few fundamental ideas (AAAS, 1993; NRC, 1996). The way that science information is organized, presented, and assessed forms the workings of effective science instruction. Many districts have designed curriculum maps that tell the teacher the what, how, when, and why of teaching science. If this map does not promote learning with understanding, framed with a series of lessons, then science is often taught in isolation.

To effectively design a conceptual unit, Larry Ainsworth and Jan Christinson (*Five Easy Steps to a Balanced Math Program*, 2006) suggest a format that sequences through planned instruction and assessment. In the field of science, additional components have been added to support the complex nature of science information, such as vocabulary, the science process skills, and literacy integration.

The next section is an overview of the steps that provide the format for designing, teaching, and assessing students' understanding of a conceptual unit.

STEPS IN DESIGNING A SCIENCE CONCEPTUAL UNIT

1. Identify the important science concepts.

    a. Identify the priority science standard(s).

    b. Think of a science concept or topic that students need to thoroughly understand at your grade level or in your particular course.

    c. Discuss your choice with colleagues.

    d. Share your thoughts with the group.

Grade-level teams should identify a particular concept or unit focus based on the standards and objectives established by the state. The topic must meet three criteria: (1) it is from the state standards; (2) it is important for the next grades; and (3) it has lifelong implications for student learning. For primary-grade students, this might include topics such as weather concepts, needs of living organisms, and concepts of sound. This topic will become the focus for the unit that will be developed and taught to students. An example from a first-grade unit (see Exhibit 3.3, in the section "Primary Example of a Conceptual Unit") illustrates some of the steps in designing a conceptual unit.

2. Create a graphic organizer. Create an outline, bulleted list, or concept map of the unit. Use as your starting point the Conceptual Unit Template (see Exhibit 3.4, in the section "Primary Example of a Conceptual Unit").

3. "Unwrap" the matching science standards to identify the key concepts and skills. "Unwrapping" the standards is the step that identifies the major concepts (nouns and noun phrases) and skills (verbs) found in the standard and objectives. There may be an integration of standards based on the topic in the unit. The "unwrapped" process requires that teachers identify the concepts from the standards and underline them. Skills (the verbs) are identified and circled. An example of an "Unwrapped" science standard is shown in Exhibit 3.1.

EXHIBIT 3.1: Kindergarten Example of an "Unwrapped" Science Standard

Note:

The skills (the verbs) are shown in bold and the concepts are italicized in this example.

Kindergarten:

The student will **investigate** and **understand** *basic needs* and *life processes of plants and animals.*

Once the standards have been "unwrapped," a graphic organizer or chart is created to show the skills and concepts in a more simplified format.

Skills	Concepts
Investigate	Basic needs of plants and animals
Investigate	Life processes of plants and animals
Understand	Basic needs of plants and animals
Understand	Life processes of plants and animals

4. Determine the science process skills that will be addressed during the unit. This step is shown in Exhibit 3.2.

    a. Find the selected science process skills that will be demonstrated during the unit.

    b. Include these on the graphic organizer.

5. Determine the priority vocabulary from the concepts. Science textbooks contain many new words that students are expected to learn and understand. Although these words represent important concepts that are essential to science understanding, they can be overwhelming for young and struggling readers. Successful instruction in science targets the essential vocabulary that is necessary for conceptual understanding but does not neglect other related words or phrases.

In your grade-level or course concepts, determine the vocabulary that must be included in the unit for student understanding. There should be five to seven key terms or phrases that support the conceptual unit.

Here are suggestions for priority vocabulary related to the kindergarten standard in Exhibit 3.1.

plant, animal, shelter, food, living, nonliving, water, air

6. Determine the Big Ideas. Big Ideas, according to Jay McTighe and Grant Wiggins in *Understanding by Design* (2005), are enduring understandings of what we want the students to retain and remember after leaving a unit. These are decided by a grade-level team that asks, "What are two or three essential big-picture understandings that we want the students to discover on their own after they complete the conceptual unit?" Big Ideas help clarify the main ideas that students need to know about the standard.

    a. Look at the priority vocabulary list to help determine the Big Ideas of the unit.

    b. Determine the essential ideas that you want your students to learn and know.

    c. Write these as Big Idea statements to guide your instruction and assessment of the unit.

For example, from the kindergarten standard in Exhibit 3.1:

EXHIBIT 3.2: Chart Showing the Skills, Concepts, and Science Process Skills

Skills	Concepts	Science Process Skills
Investigate	Basic needs of plants and animals	Observe
Investigate	Life processes of plants and animals	Communicate
		Classify
Understand	Basic needs of plants and animals	Infer
Understand	Life processes of plants and animals	Predict

- All plants and animals need food, water, and shelter.
- Living things change as they grow.
- Plants and animals live and die.

7. Write the Essential Questions. Essential Questions are written by grade-level teams of teachers to share with students at the beginning of the unit. These guiding questions are matched with the Big Ideas and represent the learning outcomes for the unit. The purpose of the Essential Questions is to help students make connections with the skills and concepts from the standard. Students should respond to the Essential Questions with their own Big Ideas after completing the unit.

The Essential Questions should drive the instructional focus of the unit. They serve as a guide for supporting what the teacher should really teach and what the students should really learn. The teacher may find that the textbook is not enough to cover the skills and concepts and must rely on additional support resources and information. As teachers begin to write Essential Questions, consider the three question-starter stems: What, How, and Why.

For example, from the kindergarten standard in Exhibit 3.1:

- What do plants and animals need to live?
- How do plants and animals change?
- Why do we need to learn about plants and animals?

8. Determine the materials, resources, and time frame needed for the unit. This will be determined by the standard. Some districts have pacing guides that drive the instructional time for the content. Materials and resources should be located prior to implementing the unit so that smooth transitions occur from day to day.

9. Decide on a discrepant event to introduce the unit. The discrepant event should be aligned with the science standard. This is always introduced prior to teaching the unit or topic.

10. Develop writing prompts. Select several writing prompts that are aligned with the standard. Depending on the grade level and the time frame for the unit, have students complete at least one or two writing prompts each week.

11. Develop a problem-solving task. The problem-solving task should align with the standard(s) for the unit. It is suggested that a problem-solving task be completed every two weeks. For example, if the conceptual unit is four weeks in length, then two problem-solving tasks would be appropriate.

12. Plan pre- and post-assessments. Once the skills, concepts, and Big Ideas have been identified, grade-level teams will design a pre- and post-assessment to administer at the end of the unit. The end-of-unit assessment should be matched to your Big Ideas and Essential Questions.

13. Design a scoring guide. Write a task-specific scoring guide to evaluate your science end-of-unit assessment. Determine the number of performance levels (three for primary).

14. Select best instructional strategies. Use your Essential Questions (which are posted in your classroom) to focus instruction and appropriate learning activities. Robert Marzano, Debra Pickering, and Jane Pollock, in *Classroom Instruction That Works* (2001), have identified nine categories of effective teaching practices. Teachers should consider which of the nine best fit the instructional techniques needed for student learning to occur during the conceptual unit.

These nine categories are:

- Identifying similarities and differences
- Summarizing and note taking
- Reinforcing effort and providing recognition
- Providing homework and practice

- Providing nonlinguistic representations

- Providing cooperative learning

- Setting objectives and providing feedback

- Generating and testing hypotheses

- Providing cues, questions, and advance organizers

15. Administer the pre-assessment.

16. Deliver unit instruction. Determine the time frame, materials needed, and instructional methodologies. If the district or school has a pacing calendar, it may serve as a guide for the time frame for the unit.

17. Administer the end-of-unit assessment. Students complete the end-of-unit assessment guided by an accompanying scoring guide, with students doing peer and self-assessments of their work. The teacher completes the final evaluation of the student work and provides feedback.

18. Complete and review the process. Take time to assess the effectiveness of your unit and allow for any corrections or adjustments.

PRIMARY EXAMPLE OF A CONCEPTUAL UNIT

Most states provide educators with a set of content standards, objectives, or benchmarks that students should know and a set of skills that they should be able to do at the end of the specified year. Many states are including a problem-solving strand that may be incorporated into the content strands or as a stand-alone standard. States are also looking at more performance-based assessments for science that integrate the problem-solving step in a conceptual unit.

Several key components of a conceptual unit incorporate problem-solving standards. A grade 1 example is presented in Exhibit 3.3 as a sample of how teachers can use the conceptual unit as a guide to select appropriate lessons, instructional materials, and learning labs and experiments. A planning template for designing a complete conceptual unit is presented in Exhibit 3.4 and is duplicated in the Appendix (Exhibit A.8) for teachers to reproduce. Specific examples for kindergarten, grade 1, and grade 2 are also found in Part 2, "Inside the Classroom."

EXHIBIT 3.3: Sample of a Completed Conceptual Unit Design—First Grade

Conceptual Unit Focus

Scientific Inquiry
Standards and indicators aligned with the unit focus
Teachers use their own state or district science standards to "unwrap" the skills and concepts

Science Process Skills:
Observing
Communication
Classifying
Measuring
Inferring
Predicting

"Unwrapped" Concepts:
Need to know about scientific inquiry

Scientific inquiry
Objects (number, shape, texture, size, color, motion)
Tools (for scientific data)
Scientific investigations
Safety procedures

Skills: Be Able to Do
Demonstrate
Compare
Classify
Sequence
Use
Carry out

Topics or Context (resources teachers will use to teach the concepts and skills):

Science textbook
Internet Web sites
Additional reading selections
Teacher lab resources

Big Ideas

1. Scientific thinking helps us solve problems.
2. Questions about the world can be answered through using our senses and appropriate tools.

Essential Questions:
1. What is problem solving?
2. How do scientists conduct investigations?
3. Why are tools important in scientific investigations?

End-of-Unit Performance-Based Assessment:

Be a Scientist
Your teacher has given you a box of materials and asks that you create an experiment to show which of the materials are magnetic and which are not.

The box contains:
A magnet
A piece of wood
A plastic soda bottle cap
A piece of wadded-up aluminum foil
A penny
A washer
A paper clip
A key
A rubber band
A nail
Another magnet

Use your data sheet to plan your experiment. Use pictures and drawings to help your understanding.

Use the problem-solving template to record your responses.

Scoring Guide

Exemplary:
All "proficient" criteria met plus:
Describes strategy used to solve the problem
Designs a classification for the objects

Proficient:
Steps to the problem are clearly outlined
Appropriate safety procedures are used at all times
Answers are correct for each item
Data sheet shows work
Lab write-up is complete
Tools are used correctly

Progressing:
Meets two of the "proficient" criteria

Beginning:
Meets fewer than two of the "proficient" criteria
Assessment task needs to be repeated

Self-Evaluation:

Teacher Evaluation:

EXHIBIT 3.4: The Conceptual Unit Template

Grade Level: **Conceptual Unit Focus:**

Standards and Indicators:
"Unwrapped" Concepts:
"Unwrapped" Skills:
Science Process Skills:
Priority Vocabulary:

EXHIBIT 3.4: The Conceptual Unit Template

Grade Level: **Conceptual Unit Focus:**

Big Ideas:

1.

2.

3.

Essential Questions:

1.

2.

3.

Resources:

(specific lessons, textbook pages, and learning activities to be used during the unit)

Materials:

Time Frame:

EXHIBIT 3.4: The Conceptual Unit Template

Grade Level: **Conceptual Unit Focus:**

Discrepant Event:

Writing Prompt(s):

Problem-Solving Task:

(specific lessons, textbook pages, and learning activities to be used during the unit)

End-of-Unit Assessment:

Selected Response:

Constructed Response:

Performance Tasks:

Instructional Strategies:

EXHIBIT 3.4: The Conceptual Unit Template

Scoring Guide:

Exemplary:

All "proficient" activities PLUS:

Proficient:

Progressing:

Beginning:

EXHIBIT 3.4: The Conceptual Unit Template

Peer Evaluation:

Self-Evaluation:

Teacher Evaluation:

Comments:

Questioning

Productive questions promote science and encourage thinking while constructing knowledge. Often, good questions have more than one correct answer.

There is no doubt that if we improve teacher questioning skills, we will improve student thinking skills. Good questions should introduce the lesson, expand on the content, and review for comprehension and understanding. Effective questions encourage students to participate in the lesson, develop critical and creative thinking skills, and become more productive test takers. Research reveals that 80 to 90 percent of the questions asked in classrooms are at the lowest cognitive level; we also know that learning to ask good questions is a process that develops over time and with practice.

BLOOM'S TAXONOMY OF COGNITIVE THINKING

To help teachers develop their questioning skills, Bloom's Taxonomy of Cognitive Development, along with strategies that form a strong, easy-to-use foundation for learning, are suggested in this section.

In 1956, Benjamin Bloom (*Taxonomy of Educational Objectives, Handbook I*) headed a group of educational psychologists who developed a classification of levels of intellectual behavior that are important in learning. Bloom found that more than 95 percent of the test questions that students encountered at that time required them to think only at the lowest possible level—the recall of information. This is also the case in many classrooms today.

Bloom's model is widely accepted today, incorporating six levels within the cognitive domain, from the knowledge and comprehension level through increasingly more complex and abstract mental levels including synthesis and evaluation.

Questioning Taxonomy

Anderson and Krathwohl (2001) revised Bloom's original taxonomy by combining the cognitive process with the knowledge components. The revised levels of thinking are: remember, understand, apply, analyze, evaluate, and create. Robert Marzano, in *Dimensions of Thinking: A Framework for Curriculum and Instruction* (1988), also addressed levels of thinking, a concept that is widely accepted and used in state test development. Bloom's original taxonomy and the revised taxonomy are depicted in Exhibit 3.5. Exhibit 3.6 shows the six categories of Bloom's cognitive verbs.

TYPES OF QUESTIONS IN SCIENCE

Higher-order thinking is one of the objectives of science teaching. All teachers use questioning in the classroom to encourage student thinking at all levels, from basic understanding to analytical and evaluative assessments. Observations of teachers indicate that most of the questions posed to students in science are at the lower level of Bloom's Taxonomy, requiring nothing more than recall of information. Many teachers consider the mastery of the text material to be their primary responsibility in getting students to answer questions. The test and state curricula often have caused a "dumbing down" of effective teacher questioning strategies. Teachers must make a conscious effort to practice and include higher-order-thinking questions into their daily lessons.

Teachers must choose the right questions, ask questions effectively, and respond appropriately to student answers. By using specific, probing questions, teachers are able to extend their own knowledge and move students into more divergent, open-ended thinking.

The following types of questions contribute to an environment of inquiry, problem solving, and higher-level thinking.

EXHIBIT 3.5: Bloom's Original Taxonomy and the Revised Taxonomy

Bloom's Original Taxonomy	Revised Bloom's Taxonomy
Evaluation	Creating
Synthesis	Evaluating
Analysis	Analyzing
Application	Applying
Comprehension	Understanding
Knowledge	Remembering

EXHIBIT 3.6: The Six Categories of Bloom's Cognitive Verbs

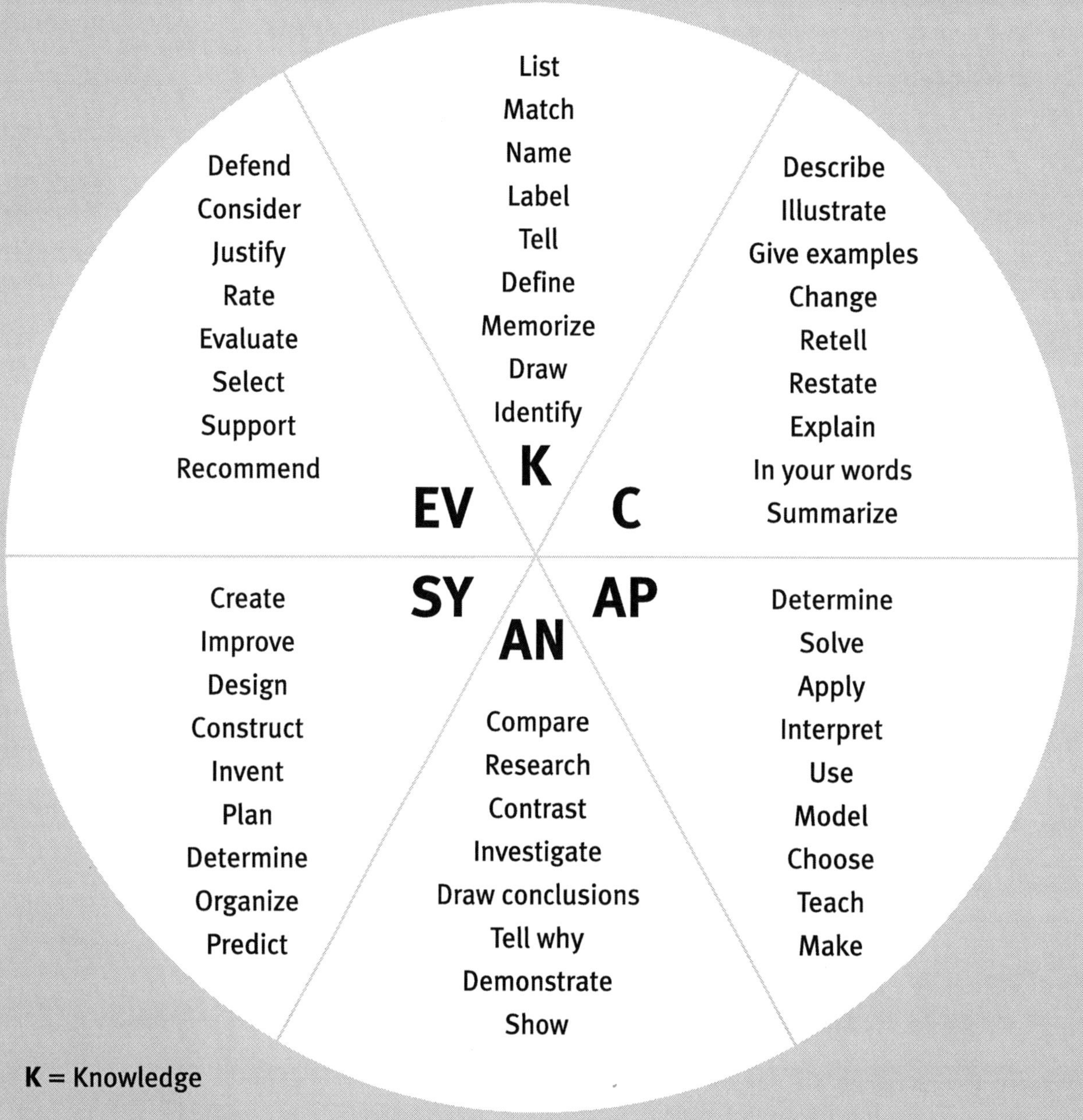

List
Match
Name
Label
Tell
Define
Memorize
Draw
Identify

K

Defend
Consider
Justify
Rate
Evaluate
Select
Support
Recommend

EV

Describe
Illustrate
Give examples
Change
Retell
Restate
Explain
In your words
Summarize

C

Create
Improve
Design
Construct
Invent
Plan
Determine
Organize
Predict

SY

AN

Compare
Research
Contrast
Investigate
Draw conclusions
Tell why
Demonstrate
Show

AP

Determine
Solve
Apply
Interpret
Use
Model
Choose
Teach
Make

K = Knowledge

C = Comprehension

AP = Application

AN = Analysis

SY = Synthesis

EV = Evaluation

1. **Interest-focusing questions.** These questions focus children's attention on using their senses and the science process skills. Teachers who are beginning a demonstration or experiment should pose this type of question for students as they explore a new topic or concept.

> Examples:
>> "How would you describe . . . ?"
>> "What did you notice . . . ?"
>> "What is it doing . . . ?"
>> "What do you think will happen . . . ?"
>> "What did you see, hear, smell . . . ?"

2. **Measurement questions.** These questions encourage better observation, classification, and communication.

> Examples:
>> "How much . . . ?"
>> "How far . . . ?"
>> "How long . . . ?"
>> "How many times . . . ?"
>> "How often . . . ?"

3. **Compare-and-contrast questions.** These questions ask children to use similarities and differences in their analysis. Compare and contrast is a process of identifying like and different characteristics. This is one of the most difficult skills for students to understand because it requires higher-level thinking, such as synthesizing and evaluating. Students are asked to determine relationships, quantify ways in which things are similar and different, and create a scheme to describe the comparisons.

> Examples:
>> "What is similar . . . ?"
>> "What do you see that is different . . . ?"
>> "How do these things go together . . . ?"
>> "What is the difference between . . . ?"

4. **Prediction questions.** These questions allow for predicting, experimenting, and investigating. This is the "What if . . . ?" question that all science teachers include in the problem-solving step because it deals with manipulating variables. This type of question requires that students use prior knowledge in order to predict an outcome of an investigation or experiment.

Examples:

"What would happen if . . . ?"

"What if you . . . ?"

5. Problem-solving questions. These questions encourage the testing of hypotheses and the formulating of conclusions. Teachers who use problem-solving questions are asking students to think at a very high level of complexity.

Examples:

"What is the best way . . . ?"

"How could you change . . . ?"

"Can you figure out how to . . . ?"

"What is another way to . . . ?"

6. Evaluative questions. These questions focus on how and why things work and on evaluating the results.

Examples:

"Explain . . ."

"Why did this happen . . . ?"

"What is the reason for . . . ?"

"Why do you think . . . ?"

"Justify your answer . . ."

STRATEGIES FOR INCORPORATING QUESTIONING

It is through questioning that students develop problem-solving skills and divergent thinking. Take time to practice developing multiple levels of questions to use with review, teacher input, and assessment. Make sure that lesson plans contain questions that are appropriately challenging for all levels of academic ability and skills. Here are a few strategies for incorporating questioning into the science classroom.

Questioning Displays

Bulletin board and classroom displays can be powerful learning tools for students. They should be simple, have a purpose, and be visible to all students. Many teachers create bulletin boards that represent specific units, and it is suggested that displays be designed to match the current conceptual unit.

Some states and districts require that teachers place the cognitive levels of thinking in a conspicuous space in the classroom. In addition, lesson plans must reflect the planning for questions during the daily instruction. A bulletin board with Bloom's Taxonomy or a version of the thinking verbs is an excellent way to have students become familiar with and learn to use the verbs. Depending on the grade level, teachers can design a visually attractive and functional display of the thinking levels.

Questioning Cubes

The questioning cubes are teacher tools to use during small-group work with science topics or content. As you work with individual groups, use the cubes to randomly "roll" a thinking-verb category. Create a question at that level of thinking and present it to the group for discussion. For example, a roll of "Comprehension" could stimulate creation of questions beginning with *describe, give examples, illustrate, or explain.* A sample template for making a questioning cube is found in the Appendix (Exhibit A.9).

Questioning Wheels

A questioning wheel may be either a teacher-directed instructional tool or a student manipulative in science. Make a transparency of the wheel and use a paper clip, in the center, to spin for the selection of question verbs for review and discussion. You may also make enough question wheels for students to independently create their own questions during small-group work and discussions. A sample template for making a questioning wheel is found in the Appendix (Exhibit A.10).

Questioning Organizers

A questioning organizer is a wonderful tool for giving notes and information to students or as an independent activity for students. Teachers should provide instruction that requires incremental thinking. Based on Bloom's Taxonomy, teachers can have students address each of the levels of thinking as the class moves through the graphic organizer. Using the Bloom's Taxonomy Wheel, an example found in a conceptual unit on simple machines is given in Exhibit 3.7. Notice how the levels of thinking increase in complexity as the activities move around the graphic. A sample template for making a questioning organizer is found in the Appendix (Exhibit A.11). The questioning organizer may be used as teacher-directed instruction or as a cooperative learning activity for students.

EXHIBIT 3.7: Example of Simple Machines Using Bloom's Taxonomy Questioning Organizer

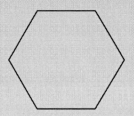

Name three simple machines.

Recommend a specific simple machine that could be used on a playground.

Describe one of the simple machines.

Design an experiment with one of the simple machines.

Choose one of the simple machines and tell how you would use it.

Demonstrate how you would use your simple machine in your house.

Cooperative Learning

Research has shown that students who work in cooperative groups do better on tests, especially with regard to reasoning and critical thinking skills, than those who do not (Johnson and Johnson, 1989). Cooperative learning allows students the opportunity to interact and engage in learning experiences. Students work together to accomplish shared goals with structures in place to support the collaborate groups.

Collaborative learning exercises enhance important skills, including (Barkley, Cross, and Major, 2005):

- Using the language of the discipline
- Explaining, providing feedback, understanding alternative perspectives
- Discovering patterns and relationships
- Organizing and synthesizing information
- Developing strategies and analysis

The key to effective cooperative learning experiences in science is the teacher's design and implementation process. It is suggested that the following steps be completed prior to assigning students to groups and roles.

1. Pre-planning. This helps to establish the specific cooperative learning technique or strategy to be used and creates the foundation for effective group roles. The teacher must decide how the groups are formed, which strategy to use, and the roles and responsibilities of each group member.

2. Introduce and model the expectations for the students. This includes explaining the academic task and the criteria for students' roles and responsibilities. Team norms, components of positive interdependence, and individual, as well as group, accountability should be fully described and explained. Time limits, use of resources and materials, and other logistics are included in this step.

3. Monitor and provide feedback. The teacher will now allow the students to work in their cooperative groups while monitoring and providing feedback to each group. Interventions for groups that are not on track is done during this step.

4. Conduct an assessment. Even though the teacher is continually monitoring and providing feedback during the students' work time, individuals and groups should be assessed at the end of the activity. Areas for assessing include:

- Focus on the assigned task
- Teamwork habits (individual and group)
- Listening, questioning, and discussing
- Research and information sharing within the group
- Problem-solving skills

5. Group processing. Each group should be required to rate their own performance within their group. Some teachers have students set goals for improving their individual and cooperative work.

Cooperative Learning Roles in Science

There are multiple roles that may be assigned to students. Students should be assigned to groups of three or four and given one of the suggested roles described below. The grade level will determine the extent to which teachers assign roles. Primary teachers will often modify and refine these roles so that their students are successful in the activity.

Note: Cooperative role worksheet templates are available in the Appendix (Exhibits A.12 through A.17).

COOPERATIVE ROLES IN SCIENCE

The following cooperative roles, described here for students, will help students understand the various functions required in problem solving.

- **Role: Creative Connector**

  Your job is to find connections between the literature and the world. This includes connecting the reading to your own life, to events at school and in the community, to outside problems, and to daily events. Make connections between this literature and other writings on the same topic or by the same author. Write your own personal connection and ask each member to make a connection. Use the response questions as a guide.

- **Role: Discussion Director**

  Your job is to develop a list of questions that your group might want to discuss about this content. Don't worry about details. Help your group discuss the main ideas and share their reactions. The best discussion comes from your own thoughts, but you may use some of the question cues to develop your discussion.

- **Role: Literary Luminary**

  Your job is to locate special sections of the text that your group would like to hear read aloud. You must help them remember some interesting, powerful, and important sections of the text. You decide which passages or paragraphs are worth reading and then plan for how they will be shared. At least one of the passages should be read aloud by you. Others may be shared reading.

- **Role: Illustrator**

  Your job is to draw a visual related to the reading. It may be a sketch, cartoon, diagram, or graphic organizer. You may draw something that's discussed or something that made a connection in the reading. You are to share your illustration before asking for group feedback.

- **Role: Summarizer**

  Your job is to prepare a brief summary of the literature read today. Make your comments one to two minutes in duration, focusing on the key concepts and ideas. List the key points. You may use any strategy for helping your group to remember the important components.

• **Role: Word Wizard**

Your job is to find and identify vocabulary words or phrases that the group finds challenging or hard to understand. After reading the text, select three to five words or phrases that need discussion. Based on the context clues, guess the meaning of the word, look it up, and be ready to share with the group.

Science Literacy

READING IN SCIENCE

Scientific literacy is the knowledge and understanding of scientific concepts and processes required for personal decision making, participation in civic and cultural affairs, and economic productivity. To be literate is to be able to read and write and show knowledge and learning.

In order for students to become scientifically literate, they must be given the opportunity to explore their interests and be curious about their world, to ask questions, to read for information, and to communicate their findings. If teachers engage students for learning and model literacy techniques and methods, students will be more receptive to the sciences.

George D. Nelson, director of Project 2061 for the AAAS, defines science literacy as:

> . . . having knowledge of certain important scientific facts, concepts, and theories; the exercise of scientific habits of mind; and an understanding of the nature of science, its connections to mathematics and technology, its impact on individuals, and its role in society.

Five Easy Steps to a Balanced Science Program looks at literacy as:

• The ability to understand and comprehend written and oral communications

• Vocabulary acquisition

• Strategies to improve reading

With the introduction of No Child Left Behind, the pressure on teachers to improve reading and math achievement became a primary goal for school systems and educators. Science took a backseat in the curriculum when extended blocks of time were allocated for reading and math. In some districts, science was not taught in elementary school, and students were not exposed to any scientific thinking or problem solving. The concern from educators was that we were neglecting a vital component in the curriculum and that we should provide methods to integrate both science and literacy.

"When learning is active, the learner is seeking something in answer to a question, information to solve a problem, or a way to do a job. Learning can't be swallowed whole. To retain what has been taught, students must chew on it."

— Mel Silberman,

Active Training: A Handbook of Techniques, Designs, Case Examples, and Tips, p. 6.

THE SCIENCE TEXTBOOK

When I first started teaching, the students were given an earth science textbook with 596 pages of text, illustrations, and graphs. The problem was that many of my eighth-grade students were not reading on grade level and I was not a reading teacher. At this point, I provided the textbook as a reference and did not rely on the text to design my instructional lessons.

Reading is essential to effective teaching and learning. The most common science reading material is the textbook. Science textbooks can be overwhelming for many students, especially those who struggle with reading. They often contain unfamiliar vocabulary and cover an extensive amount of material in each chapter. In addition, students will be exposed to laboratory manuals and teacher-prepared handouts that have different text structures than are found in fiction literature.

The science textbook has several problems associated with student reading and comprehension:

- Most tend to be encyclopedic—full of unfamiliar vocabulary and fact-heavy.
- They often include too much material to understand or cover in the given time period.
- They offer limited examples to explain science concepts.
- They don't always clearly define science concepts.
- Their reading level is often too high for the student.

To help students with reading textbooks, teachers should follow these suggestions.

1. Read the text/chapter selection prior to teaching the topic or unit. The teacher should fully understand the content of the text selection so that explanations can be given when students question the information.

2. Eliminate unnecessary sections that do not support the concept. There is too much information in the science textbook to cover in one year. If the teacher has effectively designed a conceptual unit, then only the text sections that support the standard should be addressed during the lesson. This is not to say that the other information is not valid; rather, it can be explored through extension activities.

3. Focus students on the concept that will be taught. This comes from having identified the "unwrapped" skills and concepts from the standards.

4. Identify difficult vocabulary and develop a plan for understanding the terminology. Science textbooks are overwhelmingly difficult for some students due to the complexity of the vocabulary. Teachers should present the new terms prior to having students read the text.

5. Give relevant examples to support the concept. Accessing prior knowledge makes connections for students in the form of text to self, text to text, and text to the world. For example, if students have not been exposed to different types of trees, based on where they live, they may not have prior knowledge of conifer or deciduous trees.

6. Prepare questions before teaching the chapter or unit. All lesson plans should contain the guiding questions that will be used to drive the instruction and assessment of the lesson. Bloom's Taxonomy of Cognitive Thinking is an excellent tool for formulating questions ranging from a low level to a high level of thought.

7. Read and chunk the text information by using effective note-taking and summarization strategies. Most science textbooks are designed in "chunks" of information and small sections. After a section of the chapter is read, the teacher should model and have students practice finding the main idea and key information. Graphic organizers for note taking and summarizing are useful tools for students to represent their learning.

UNDERSTANDING INFORMATIONAL TEXT

Teachers must provide explicit instructions on how to use and read expository/informational text. When given the opportunity, young children can listen to, read, and write informational text to build an understanding of the Big

Ideas and important concepts in science. Most state reading standards and assessments expect that children can read and write informational text by fourth grade or earlier.

For example, the 2003 Arizona Reading Academic Standards for Kindergarten state:

Strand 3: Comprehending Informational Text

Comprehending Informational Text delineates specific and unique skills that are required to understand the wide array of informational text that is part of our day-to day experiences.

Concept 1: Expository Text – Identify, analyze, and apply knowledge of the purpose, structures, and elements of expository text.

PO 1. Identify the purpose for reading expository text.

PO 2. Restate facts from listening to expository text.

PO 3. Respond appropriately to questions based on facts in expository text, heard or read.

Science content learning in the early grades is often an interdisciplinary approach combining science with reading, math, art, music, and drama. The challenge for primary teachers is to develop a balance between the content-specific units and best instructional practices for early readers. Primary teachers must connect science information in expository text to real-life experiences in children's lives. Informational text contains contents that can be fun, interesting, and motivating for children.

Students should be shown how to look at the text for informational purposes by having them examine the various types of text structures within nonfiction literature selections used in the classroom. Proficient readers learn how to understand the features. When students explore and understand each of the text components, they are able to make meaning from the information.

Duke (2004) recommends four strategies to help teachers improve student comprehension of informational text. Teachers should:

• Increase students' access to informational text

• Increase the time students spend working with informational text

• Teach comprehension strategies through direct instruction

• Create opportunities for students to use informational text

Examples of features found in nonfiction text include:

• Table of Contents

• Appendix

• Index

• Bold print

• Italics

• Diagrams

• Pictures

• Captions

• Glossary

• Charts

READING ALOUD

Pairing nonfiction and fiction books to read aloud is an effective way to present science content to students. Fiction books can be used as the "hook" that gets students interested in the content. Nonfiction books and other informational texts should also be presented.

Today, there is an abundance of science information, both in printed format and on the Internet. Literature provides students with a context for the science concepts explored in and out of class.

In 1985, the Commission on Reading report, *Becoming a Nation of Readers*, included among its findings that "the single most important activity for building the knowledge required for eventual success in reading is reading aloud to children" (Anderson et al.).

The commission backed up its conclusion with research that indicated that reading aloud in the home is an essential contributor to reading success, and that reading aloud in the classroom is "a practice that should continue throughout the grades."

Being read to creates interest in reading and literature, improves vocabulary skills, builds upon background knowledge, and fine-tunes listening skills. Reading aloud is appropriate for all grade levels and subjects.

Teachers play a major role in creating a learning environment that nurtures reading development. We model our love of reading by reading aloud to students. When your students hear you model all of the elements of good reading, they realize that reading is more than word calling, and they are likely to become more willing to learn the skills necessary to read independently with expression. Reading aloud daily is possible with purposeful planning. Read your favorite passages of a story or play, trivia or adventure stories, interesting and weird facts, and newspaper clippings.

Most elementary school teachers are well versed in language arts and the integration of reading and writing in class. Elementary classrooms around the country have a designated block of time for literacy, and teachers utilize the best practices of teaching reading. When children are allowed to use their own curiosity with a science reading selection, their natural curiosity about the world around them is presented in books with vivid text and beautiful illustrations.

As a science teacher, you have the responsibility of creating a science-literature-rich environment for your students. Textbooks should not be the only reading resource in the classroom. A well-designed science classroom has a variety of reading material for students, including texts at different reading levels, a variety of science journals and magazines (e.g., *National Geographic for Kids, Sierra, Ranger Rick*), and science reference books. Technology must be available for students to access the Internet and connect with scientists around the world. Science teachers should also request that the school media center be current in the latest grade-level-appropriate nonfiction and fiction science resources.

READING STRATEGIES FOR THE SCIENCE CLASSROOM

As teachers, we have the responsibility to incorporate reading and writing into our daily lesson plans. It is imperative that all teachers become teachers of reading. By using a variety of before-, during-, and after-reading strategies, students are given opportunities to become better readers.

Romance and Vitale (1992) found significant improvement in both reading and science scores of fourth graders when the regular basal reading program was replaced with reading in science that correlated with the science standards. Students' attitudes toward science also improved. However, the students' attitudes toward reading in general stayed about the same.

One of the most effective strategies to get students interested in reading is to introduce a new book each week. You can actually "sell" the book to the class by showing the cover, telling students about the book, and demonstrating enthusiasm about the book subject. Students are given the chance to read the book during the week and share their thoughts.

STAGES OF READING

The strategies that are included in *Five Easy Steps to A Balanced Science Program* are appropriate for all grade levels and content areas. Teachers in the early grades will need to modify the implementation of the strategies to meet grade-level needs of their students. For example, a kindergarten teacher will use more vocalization than linguistic delivery of the strategy. As students develop proficiency in writing and reading, more emphasis should be placed on these skills.

The stages of reading are:

- **Pre-reading:** Activating prior knowledge
- **During reading:** Understanding and comprehension
- **After reading:** Synthesizing information

> *"Desultory reading is delightful, but to be beneficial,*
> *our reading must be carefully directed."*
>
> —ANNAEUS LUCIUS SENECA, ROMAN PHILOSOPHER (4 BCE–65 CE)

Pre-reading

Fluent readers use a variety of strategies to comprehend text selections. Pre-reading strategies are those that help students build understanding about a text before they actually begin reading. This is what students actually do with the text. When the teacher effectively models these strategies, students develop a deeper understanding of the science content. It is important that the teacher identify text that is directly related to the conceptual unit standard and corresponding topic. Readers will pay more attention to what they are reading if they can relate the text to their previous experiences. Part of the comprehension and meaning difficulty that occurs when reading informational text comes from a lack of prior knowledge. Using pre-reading strategies as a teacher-directed method helps students build background knowledge when little or none exists.

Each of the following pre-reading strategies is based on best instructional practices in literacy. Select the method that meets the needs of your students. Some work prior to the students seeing the strategy is required in each of the activities.

T.I.P. = Title, Information, and Preview

The purpose of T.I.P. is to introduce students to a new text selection prior to reading the information. This strategy is teacher directed and can be used with any new text selection, chapter, or book. Prior to reading the selection, the teacher asks the students to name the **title** of the text. The students then have a chance to look through the text to determine what kind of **information** they will be learning as they read. Guiding questions such as "What do you think we will learn?" and "What do you think the author wants us to know?" help frame students' understanding. Once students have had time to answer these questions, they should **preview** the selection for visuals, charts, graphs, diagrams, and illustrations. These are the informational text structures used in scientific writing. This is a powerful test-taking skill in that students are directed to view the information in the text selection prior to reading. Many students do not pay attention to the ancillary materials found in text reading material, and this strategy will help them focus on information that may support the test questions.

Tell Me What You Know

The purpose of Tell Me What You Know is to access prior knowledge before reading or exploring a science topic. The graphic organizer is used to gather students' comments and knowledge about a topic. Exhibit 3.8 provides a good structure for a graphic organizer. The teacher must guide the discussion based on what is most relevant in the text. Student comments are written on the graphic organizer for all to see. About five to ten minutes are spent with students sharing what they know. Guiding questions such as "How do you know?" and "Why do you think so?" give the teacher an insight into student learning. Once students have shared their comments, the teacher instructs them to find their answers within the text and confirm their learning. Wrong information will be given, but this is an opportunity to show students that their thinking is not always correct. If the text allows, students may highlight their newly found information. At this point, the topic is introduced and fully discussed.

Anticipation Guides

The use of anticipation guides is a pre-reading strategy that allows students to access prior learning through a decision-making process. After selecting the

EXHIBIT 3.8: Graphic Organizer for Tell Me What You Know

Tell me what you know:

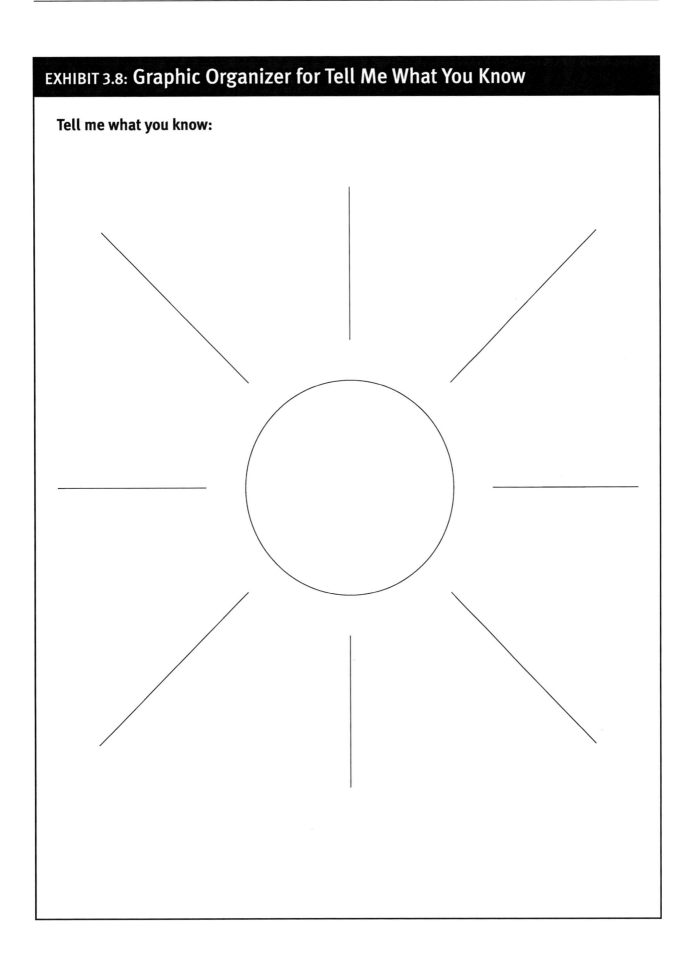

relevant text selection, the teacher develops four to six statements, some of which are true and some of which are false. These are presented to the students, who must decide which are correct statements and which are not. Students can do this activity with a partner or in a cooperative group. Once answers are shared with the class and teacher, the students read the text selection to find the answers to the statements. It is recommended that students be held accountable for the location, within the text, of the answers.

A first-grade unit that includes the desert habitat might use the following statements as an introductory anticipation guide.

____ A desert is a place that receives less than ten inches of rain per year.

____ The Sahara is the largest desert in the world.

____ There are no deserts in North America.

____ Camels are often called "the ships of the desert" because they can cross the desert better than any other animal.

____ Some deserts have no sand dunes.

Word Splatter

Word Splatter is a high-level thinking strategy used with pairs of students or cooperative groups. Prior to reading the text, students are presented with a list of six to eight key words, phrases, and numbers related to the content of a text selection. These are "splattered" on the board in random positions. Students must decide how the words, phrases, or numbers are relevant to the text. Time is provided for students to share with and justify their explanation to others. The text is then read as students find the information within the selection.

A kindergarten unit on the human body could include these words:

bones	legs	arms	head
feet	nose	toes	fingers

The P.O.A. Method (People, Objects, Action)

Students today are very visually oriented, and actual photos of people exploring and studying science provide additional opportunities for students to

EXHIBIT 3.9: Questions to Help Guide Discussion on Science Topics

Essential Questions	Evidence to Support Your Thinking
Are there any people in the picture?	
Are there any objects in the picture?	
Is there any action in the picture?	
If there is an object in the picture, what do you think it is?	
How big do you think it is?	
Where can you find it?	
What do you think it does?	
Why is it important?	

discuss what they know about a topic before it is investigated. After the teacher identifies the standard and topic, a picture related to the topic is shown to the class. For example, a photo of a volcanologist near a lava flow would bring interest to the unit prior to instruction. Students are held accountable when the teacher requires them to respond using details and information found in the picture. The questions in Exhibit 3.9 may help guide the discussion. Additional questions may be developed by the teacher to specifically relate to the standard or topic.

During Reading

Effective readers, when they are proceeding through a text selection, continually monitor their comprehension to make sure that the information makes sense. Students create understanding and comprehension during the reading of a scientific text selection. Teachers can help students with their reading by using the following four strategies.

The K.I.D. Method (Key Vocabulary, Information, Details)

The K.I.D. Method enables the students to identify three sources of information that will help them to better understand the text. The **key vocabulary** within the text is identified, discussed, and recorded on the template (Exhibit 3.10) in their notebooks. For example, if the article is about volcanoes, key vocabulary might include *lava*, *magma*, *volcanologist*, and *ring of fire*. Important **information** about the selection is recorded in their notebooks for discussion and sharing. Supporting **details** about the article include any interpretation related to charts, graphs, diagrams, maps, or illustrations. It is suggested that, once this class activity ends, the students do a quick write-up or summary statement of what was learned.

Five–Three–One

Five–Three–One is a strategy that focuses on identifying the author's purpose and the main idea of a science text selection. Students may work individually or in pairs to determine the key facts or concepts in the text. The strategy involves the students narrowing their findings down to the one best explanation that summarizes the reading. A template for Five–Three–One is provided in Exhibit 3.11.

EXHIBIT 3.10: Template for the K.I.D. Method

Text Selection:

K **Key Vocabulary**	I **Information**	D **Details**

EXHIBIT 3.11: Template for the Five–Three–One Strategy

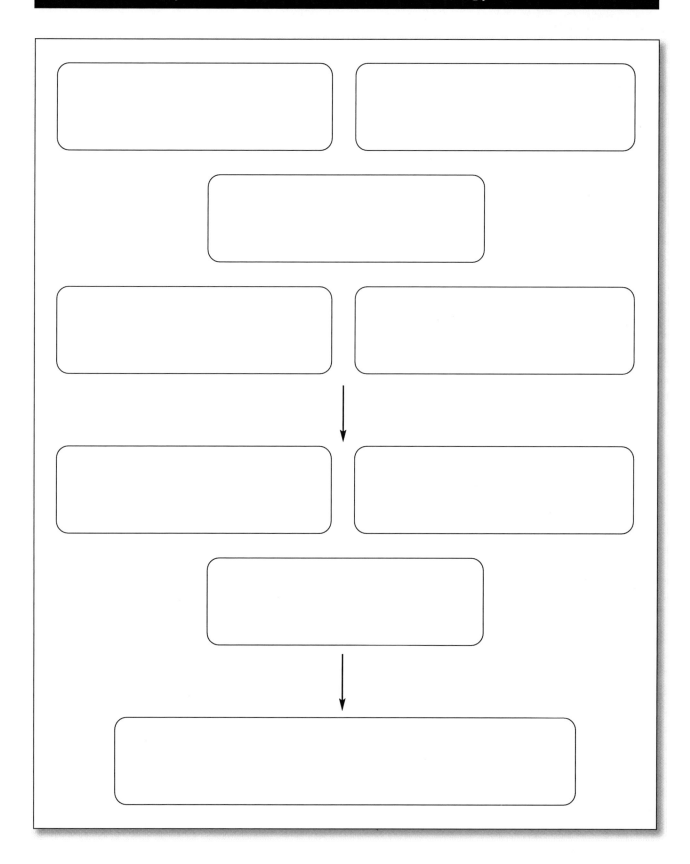

Four-Square Thinking

Four-Square Thinking is a cooperative learning strategy that provides a method for sharing science information from a text selection. Students are numbered off in their group from one to four. Each student reads the text section as instructed by the teacher. Using the Four-Square Thinking template (Exhibit 3.12), each student writes, illustrates, or records his or her summary of the text in one of the blocks. The teacher may want the students of like numbers (i.e., all of the ones, all of the twos, etc.) to meet and discuss their responses. This strategy may also be done kinetically, with students getting up and forming a square for discussion. A discussion within each group occurs before comments are shared with the whole class.

Read–Write–Share

Read–Write–Share is modeled after the "quick write" process. Students read the science text selection and then write a quick summary of what they read. This will focus their learning on only the key information in the reading. Have students share with a partner and then with the teacher.

After Reading

Once students have read a scientific text selection, they must synthesize the information into their own words to deepen understanding of the content. Effective readers use behaviors that allow for closure to the text. These include retelling, summarizing, and evaluating. Strategic readers personalize and internalize the meaning of the text.

Three–Two–One

Three–Two–One is a summarizing strategy that reviews the important information read from a text selection. Ask students to respond to the following three statements (which may be changed by the teacher):

List the 3 most important things you learned.

Write 2 questions you still have about the content.

Share 1 strategy that will help you remember this material.

ABCD Assessment

The ABCD assessment strategy is appropriate to use at the end of the day, at the end of a unit, or as a transitional activity. The purpose is to allow students a

EXHIBIT 3.12: Template for the Four–Square Thinking Strategy

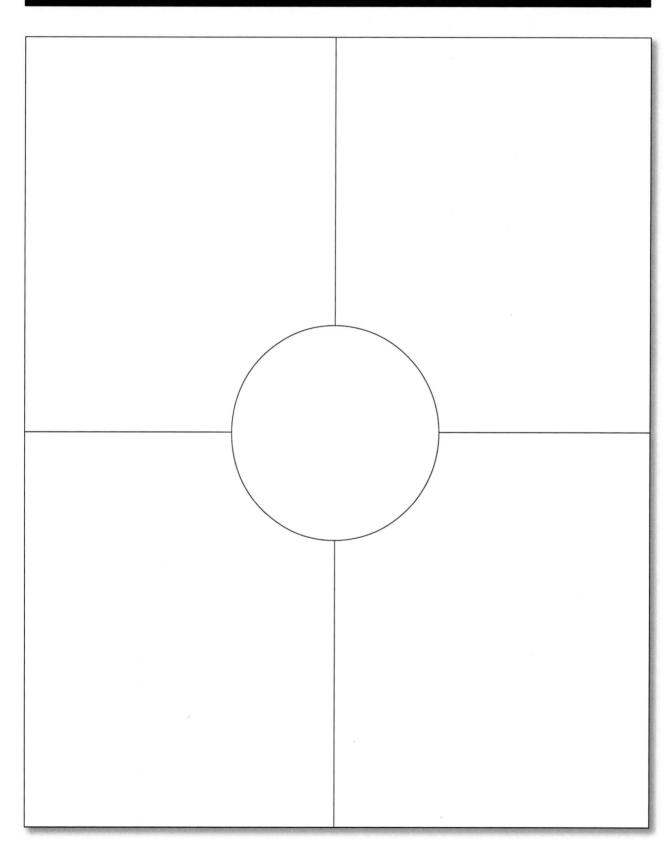

chance to practice their test-taking skills using multiple-choice questions. The teacher creates a series of questions related to the conceptual unit or science topic and presents these to the students. Using cards with the letters A, B, C, and D, students hold up the correct response to the asked questions. Elementary students may use fingers in place of cards.

My Summary

Students who have an opportunity to self-reflect will internalize the content faster than those who are not involved in personal assessments. My Summary targets student thinking using only 10 words as an indicator of learning. After reading the material, students should summarize the information in fewer than 10 words. This will be difficult in the beginning, so allow time for modeling, practice, and sharing. Starter statements such as "I learned . . . ," "The most important things were . . . ," or "I thought . . ." may be given to students as a learning tool for getting started with this strategy.

Reflecting on My Learning

- What is the most important thing to remember in Step 3?

- What strategies will I try with my students?

- What additional information do I need?

STEP 4: Mastery of Science Information

CHAPTER 4

S TEP 4 PROVIDES THE BEST INSTRUCTIONAL PRACTICES for engaging students in science knowledge and skills. The focus is weekly science review and writing in the sciences.

Essential Question

How do students learn science?

✎ KEY POINTS

1. Conceptual understanding in science incorporates thinking, reasoning, and making connections between specific content and real-world applications.

2. Writing in science improves reading in science.

3. A strong understanding of science terminology and vocabulary enables students to comprehend complex scientific information.

4. Nonlinguistic representations are tools that students can use to organize their scientific thinking and problem solving.

HOW PEOPLE LEARN

Inquiry and the National Science Education Standards (National Research Council, 2000a) summarized the findings of *How People Learn* (Bransford, Brown, and Cocking, 1999), which support the use of inquiry-based learning. The findings include these points:

- Understanding science is more than knowing facts.

- Students build new knowledge and understanding on what they already know and believe.

- Students formulate new knowledge by modifying and refining their current concepts and by adding new concepts to what they already know.

- Learning is mediated by the social environment in which learners interact with others.

- Effective learning requires that students take control of their own learning.

- The ability to apply knowledge to novel situations is affected by the degree to which students learn with understanding.

"Science is facts; just as houses are made of stones, so is science made of facts; but a pile of stones is not a house and a collection of facts is not necessarily science."

—Henri Poincare, French mathematician and physicist (1854–1912)

THE WEEKLY SCIENCE REVIEW

How many times have you thought that your students' previous teacher "just didn't teach anything?" For years, I could not understand how my students could forget so quickly from year to year. This was especially disconcerting when I moved up with the kids and had the same ones from the previous year. There was no one to blame but myself.

There is so much that teachers have to address with science content standards. Some elementary teachers have the daunting task of teaching everything within the span of a school day. Sometimes science is given a very short period of time during the week. Elementary school teachers who are in departments often have more time for science instruction.

The weekly science review is designed to focus on the key concepts and understandings that are important for science skill and knowledge acquisition.

The value of the review is that teachers can use it to identify science misconceptions, missing vocabulary comprehension, and problem-solving skills. This allows students to see that science is not just a collection of facts, but rather interrelated Big Ideas and concepts.

The purpose of the weekly science review is not to instruct students on new information, but to review and practice prior learning. It is suggested that this step be placed at the beginning of the science block.

The Weekly Science Review Templates

Five Easy Steps to a Balanced Science Program provides three different templates for the weekly science review.

The **content-based template** (Exhibit 4.1) is designed for teachers to use for problems and questions related to specific science content areas by grade level. There are six boxes in this review. Teachers choose the content areas. For example, life, earth, and physical sciences are found on most state standards and would probably be represented in the weekly review. A vocabulary box or a science process skill question may be added to the review. It is entirely up to the teacher to determine the best categories for the science review.

The **standards-based template** (Exhibit 4.2) is designed to focus on the science standard being taught during the conceptual unit. Teachers may choose to include problems based on the Essential Questions, Big Ideas, priority vocabulary, or a science process skill. It is suggested that there always be a question or problem that relates the content to the real world of science. Many teachers around the country have told me that measurement and graphing are areas in which students need additional work. The template in Exhibit 4.2 is a good place for these questions or problems.

The **taxonomy-based template** (Exhibit 4.3) encourages the development of higher-order thinking skills related to the standards in the conceptual unit. Based on Bloom's Taxonomy, problems and questions are created for increasing complexity and thought. Question statements beginning with Bloom's verbs will help teachers design appropriate levels of questions for each box.

Depending on the grade level, it is suggested that teachers present the weekly science "boxes" at their discretion. For example, a kindergarten teacher may give the students only two of the boxes and a second-grade teacher may present three or four. The purpose of the weekly science review is not to

EXHIBIT 4.1: Content-Based Science Review Template

Earth and Space Science	Physical Science
Environmental Science	Life Science
Science Process Skill	Vocabulary

EXHIBIT 4.2: Standards-Based Science Review Template

Big Idea	Content/Concept
Vocabulary	**Science in the World**
Science Process Skill	**Measurement/Graphing**

EXHIBIT 4.3: Taxonomy-Based Science Review Template

Knowledge	Comprehension
Application	**Analysis**
Synthesis	**Evaluation**

instruct students on new information but to "review" and practice prior learning. It is suggested that this step be placed at the beginning of the science block.

Areas to Address in the Weekly Science Review

There is a process for selecting science review questions and problems. Individual teachers or teams choose specific topics representing the science standards for their grade level. Some suggestions are provided here as guidance in designing the weekly science review.

- **Select the format for the weekly science review.** Three types of weekly science review formats are presented in this book. One is a standards-based template that directly correlates with the conceptual unit. The second is content-based and corresponds to the grade-level state standards that are either tested or used for accountability purposes. And the third is a question-based review template. No matter which template is used, only the Priority Standards should be represented in the weekly science review.

- **Make sure that the reviews reflect science standards for a specific grade.** Teachers must determine the science standards or strands for their specific grade level. This becomes the focus for the questions or problems.

- **Match the conceptual focus to the current instructional unit.** The problems or questions on the weekly science review should have a direct correlation with the unit being taught. For example, a first-grade unit on sound would have the corresponding boxes aligned with this topic. If the content-based template is used, then life, earth, physical, and environmental sciences, plus vocabulary, would correspond to the topic of sound.

- **Reinforce prior learning.** The weekly science review is an opportunity for students to share knowledge learned during the year. It reinforces skills and content that have been taught and are now being reviewed.

- **Provide weekly practice for the state test.** The format of the questions and problems in the weekly science review should mirror the design and framework of the state test. This provides students with practice for the test and makes them familiar with the types of questions, the test design, and the skills needed for taking tests. This weekly practice may include multiple-choice-type questions, interpretation of charts and graphs, or problem-solving questions.

• **Promote scientific reasoning and processing.** At least one of the problems in the weekly science review should focus on scientific reasoning, thinking, and the process skills. It may be a short, constructed response in which students must determine the outcome of an experiment.

Resources for the Weekly Science Review

There is a wealth of information on science, both in textual format and on the Internet. Teachers should plan time for selecting items based on their conceptual unit and on the specific standards. Some suggested source include:

• Science texts

• Teacher-designed items

• Released items from local and state tests

• Science Web sites (see the Webography at the end of the book)

Methods for Processing Science Review Problems

For kindergarten teachers, I suggest that the process begin as teacher-directed instruction and be delivered through whole-group time. The teacher could present one, two, or three of the boxes to the students and have students respond verbally, using manipulatives, illustrating, or writing. This would depend on their level of ability when entering school and throughout the year.

First- and second-grade teachers must also begin with teacher-directed instruction but will transition as the year progresses. These students can write their answers but may also use other modalities to answer the questions. Once students understand the weekly science review process, the teacher could transition to having students work with a partner. Moving toward cooperative groups and finally into independent work is the goal for the weekly science process.

The Teacher's Role

The teacher is responsible for designing, implementing, and holding students accountable for the weekly science review. One of the best ways to begin implementing the review is to use it as a "warm-up" or opening science class activity. Students must be taught the procedures for beginning the review, what to do while they are working, and how they will be held accountable for the work. The teacher would circulate through the room, if not doing direct instruction, to monitor and support the students as they work. The suggested time frame is ten to fifteen minutes, but this may be adjusted based on the daily schedule.

Once students finish the problems, there are a variety of instructional strategies for processing the work. One of the key factors with the weekly science review is that it is a time for dignifying correct and incorrect responses. Mistakes are perceived as chances to improve. Whether or not the teacher wants to collect a grade for this activity is determined by the procedures governing the school or district.

Holding Students Accountable

Once students have learned the weekly science review process and procedures, teachers may want to vary the way they check for understanding. What is most important is that students identify where their mistakes are made and have an opportunity to correct any errors in their thinking. Here are a few engaging strategies from which teachers may choose for reviewing student responses.

1. **Mix and match.** The teacher selects one of the problems that the majority of the students are not understanding and addresses this particular one.

2. **Peer checking.** Students are paired together and review the problems and questions.

3. **Board work.** The teacher selects one student from each group to go to the board to share his or her answers with the class.

Suggested Time Frame for Implementation

Classroom science teachers are busy people and often say that they don't have time for "one more thing." The weekly science review is a powerful tool that does allow students to continually see and process information that has been taught.

For elementary science classrooms, teachers who have a limited amount of time for science (i.e., a forty-five-minute block or less) may want to do only one or two boxes a day. The selection of the type of weekly science review is left to the discretion of the teacher and should be based on a data analysis of what the students need to master in a particular area. This could be a combination of the three templates or a concentration on just one. Exhibit 4.4 provides an example of a standards-based weekly science review; Exhibit 4.5 shows an example of a content-based weekly science review; and Exhibit 4.6 shows an example of a taxonomy-based weekly science review.

EXHIBIT 4.4: Example of a Standards-Based Weekly Science Review

Day of Week	Weekly Science Review Boxes
Monday	Big Idea
Tuesday	Content/Concept
Wednesday	Vocabulary
Thursday	Science in the World
Friday	Using Science Process Skills

EXHIBIT 4.5: Example of a Content-Based Weekly Science Review

Day of Week	Weekly Science Review Boxes
Monday	Life Science
Tuesday	Earth/Environmental Science
Wednesday	Physical Science
Thursday	Vocabulary
Friday	Using Science Process Skills

EXHIBIT 4.6: Example of a Taxonomy-Based Weekly Science Review

Day of Week	Weekly Science Review Boxes
Monday	What would happen if . . . ?
Tuesday	What solutions might you add to . . . ?
Wednesday	Why do you think . . . ?
Thursday	Can you explain what is happening . . . ?
Friday	What examples can you find to . . . ?

WRITING IN SCIENCE

Science teachers are often not reading- and writing-specific teachers. Writing in science will often take a backseat in the list of standards, content, and assessments that are required of all teachers. Writing in science can help students understand the basic science concepts, appropriately use scientific vocabulary, and expand their science knowledge. Students who write in science class improve critical thinking, place more focus on the science process skills, and develop a deeper conceptual understanding of science content.

Writing must be an active process in science class. As students write, they must organize information, communicate their findings, and think at a higher level. The writing process enables students to understand, in their own words, difficult science content and vocabulary. When teachers provide appropriate feedback and constructive criticism, student writing skills improve along with their scientific knowledge.

Four Fundamental Reasons Why Every Student Should Write Frequently*

1. Writing improves reading comprehension.

2. Writing improves student performance in other academic areas, including social studies, science, and mathematics.

3. Writing contributes to a sense of connection and personal efficacy by students' participation in society.

4. Writing, particularly with evaluation, editing, revision, and rewriting, will improve the ability of a student to communicate and will lead to success on state and local writing tests.

The conceptual unit provides opportunities for teachers to integrate reading and writing on a regular and consistent basis. Whether reading aloud or silently, students should have the chance to read what has been written and to actively write about what they have read.

Four Basic Types of Writing

While many science teachers typically think of the lab report as the primary format for science writing, there are several ways for teachers to integrate

*Source: Douglas Reeves, from the Introduction in *Write to Know Series: Nonfiction Writing Prompts for Science*, Michelle Le Patner (2005), p. 28.

reading and writing in science. Students should be exposed to four types of science writing during the year, as depicted in Exhibit 4.7.

Narrative writing is a story that describes a sequence of fiction or fiction-nonfiction events and experiences. This type of writing typically is represented by novels, short stories, biographies, autobiographies, historical accounts, poems, plays, and essays. Narrative writing plays an important role in the science learning experience for students. Elementary teachers will begin a unit with a read-aloud from a narrative story. Examples of such stories include *Cloudy with a Chance of Meatballs, The Hungry Caterpillar,* and *Mouse Paint.*

Expository writing is designed to inform, explain, clarify, define, or instruct the reader on a specific topic. This type of writing is found in all nonfiction text examples, including newspapers, magazines, pamphlets, reports, research papers, and scientific literature. Expository writing requires that the writer give information or explain a topic. This writing is factual and comprises most of the text in science textbooks. The science process skills are illustrated through expository writing in the form of classification, comparison, and cause and effect. Nonfiction children's literature about science is a teacher tool for introducing a unit or topic to the students.

Persuasive writing attempts to convince the reader of a particular point of view or opinion. Speeches, editorials, and advertisements are examples. The writer asks that the person reading the text take a position either for or against the information. Persuasive writing is all around and is used by scientists as part of the scientific method when results and conclusions are presented at the end of an experiment or lab.

Descriptive writing relies on concrete, sensory detail to explain or communicate the point of the text. This writing appears almost everywhere and is often combined with another genre of writing. It uses senses to portray a person, place, or thing so that the reader is able to visualize the writer's words. There is always attention to detail in descriptive writing. When descriptive writing is used as a prompt, the word "describe" often begins the statement. For example:

Describe a storm.   Describe a forest.

Describe a beach.   Describe a fish.

One strategy that science teachers use for descriptive writing is the use of photographs that relate to the conceptual unit topic. For example, the unit on ecology may begin with the teacher showing a photograph of a rainforest and

EXHIBIT 4.7: Four Types of Science Writing

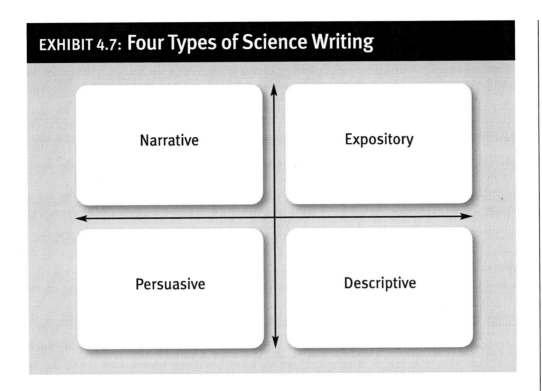

having students brainstorm words related to the picture. The students could then write about a rainforest using the descriptive words.

Writing Prompts

Writing provides an excellent opportunity for students to show what they have learned about science concepts. It is one of the most powerful instructional strategies that educators can use to engage students in active learning. Teachers have the opportunity to design and implement a variety of writing prompts based on the conceptual unit and the standards. As teachers design specific writing prompts to match the conceptual unit, there must be a conscious effort to design effective prompts that motivate and encourage writing.

How to Use Writing Prompts

Teachers may use the science writing prompts in a variety of ways during the year. Based on the type of prompt, teachers have the choice of when to ask students to write in science. These prompts can be given during or at the completion of a unit of study. Science writing prompts are not designed as the sort of "cold" prompts that are typically found in language arts class. The purpose of using writing prompts is to give students practice with all four types of writing, which will improve their comprehension of the science concepts they're writing about.

When to Use Writing Prompts

Teachers who use writing prompts have multiple options for when and how to use them with science instruction. A writing prompt can be used:

1. As a warm-up or bell activity before class starts. Students should keep their writing in their science journal. If time permits, the teacher could ask students to share their writing, but only if students feel comfortable with this activity.

2. As a quick writing activity after a reading selection has been completed. Students would take up to three minutes to write their thoughts about the text.

3. As a performance assessment in which students demonstrate what they have learned about a concept or topic.

4. As part of an interactive journal after a lab investigation or experiment.

5. As a cooperative learning activity in which students rotate a prompt around a group of four, round-robin style. Each student would have an opportunity to respond to the prompt and share with his or her classmates.

6. As a ticket out the door, with students handing the teacher their responses to the writing prompt upon leaving the classroom.

7. With a photo or image related to the topic or content.

8. With a word bank for students who need additional vocabulary support.

How to Design Writing Prompts

Teachers should design writing prompts that:

• Are relevant to the students and worth writing about

• Are aligned with the current conceptual unit of study

• Are simple and not time-consuming

• Can be answered briefly (usually in two to four sentences)

• Can be written with correct and appropriate language conventions (e.g., spelling, capitalization, and punctuation)

Types of Writing Prompts

ATTITUDINAL. Attitudinal writing prompts focus on students' feelings about themselves. The responses to the prompt provide the teacher with insight into students' preferences regarding learning and thinking.

CONTENT. Content writing prompts focus on the science Priority Standards and objectives. These are specifically designed so that students demonstrate mastery of science content.

PROCESS. Process writing prompts allow students to show understanding of problem-solving and thinking levels. These prompts allow the students to show mastery of the science process skills through the problem-solving model.

Science Writing Prompts

Primary students write best about what they know. This includes themselves, their home and family, places they have been, and their senses. Early elementary students will use visual aids in their writing, which can take the form of drawings, paintings, cut-out pictures, computer clip art, or personal pictures. Teachers can also use photos and science images to stimulate student writing.

In the strategy called People, Objects, Action (P.O.A.), students are presented with a photo of someone doing science. For example, a picture of a volcanologist near a lava flow would be used in the unit on landforms. Students are asked to identify any people in the photo and hypothesize about what they might be doing. They are to tell about any objects and describe the action in the picture. After a quick discussion, students write one to three sentences summing up what they know about the image.

In another example, a teacher shows a picture of a safari hat and asks students to write about what the person would be doing if he or her had to wear this type of hat. Teachers should see students using the science process skills to respond to this type of prompt.

As you design science writing prompts, keep in mind the types of writing (i.e., expository, descriptive, narrative, and persuasive). Another way to consider writing prompts is to focus on content (actual science information), process (thinking about science), and product (experiment results).

Although this list is not inclusive in any way, here are some primary science writing prompt samples.

"If you were a plant, what kind of plant would you be and why?"

"Suppose someone has never seen a thunderstorm. What would you tell them about this type of weather occurrence?"

"What is your favorite place to go? Use your senses to describe this place."

"Explain how you used the science process skills this week."

"Use the word 'rainforest' in a sentence. Then, write one or two sentences that show you understand the meaning of rainforest."

"Describe a practical use of one of the simple machines we studied."

"If science was a color, what color would it be and why?"

"My favorite thing about science is . . ."

"Draw a picture of yourself 'doing' science."

"How would you describe a (insert word here)?"

"The most important thing you learned about is . . ."

"Describe a practical use for a (insert word here)."

"Compare and contrast a conifer tree and a deciduous tree."

"Write a story about your favorite weather. Think about the weather and what you might wear to go outside. Think about some fun activity that you could do in this type of weather. Draw a picture to go with your story."

"You have to tell a friend how to build a terrarium. Include your steps and the material you will need. You may use a drawing with labels."

Science Journaling

What Is a Science Journal?

The science journal is a compilation of the activities, notes, labs, responses, and key information that students collect during the course of a unit or grading period. The journal is a method that allows students to keep a written and visual account of what they do and learn during science experiences. The journal helps students reason, write about, and illustrate their inquiry-based learning; develop an in-depth understanding of science concepts; and respond to teacher assessment and self-assessment. Journals are continually updated during the conceptual unit, and student entries will vary according to individual learning styles and modalities.

A teacher may use the science journal as a source for modeling the structures of nonfiction text. Included in the journal is a table of contents and often a science glossary or "Gems of Wisdom" page with key vocabulary and terms.

Within the pages of the journal, students may incorporate charts, graphs, drawings, and diagrams. Each of these would have a title, headings, labels, and captions that represent nonfiction text structures.

The science journal enables students to make information come alive using words and illustrations in age-appropriate ways. Both literacy and science process skills are emphasized as students move from day to day in their journal. During the course of a unit or grading period, the journal continually changes and expands with information that is presented in class and completed outside of class.

Through many years of experience with science teachers and all levels of students, I have developed several recommendations that have been useful for implementing science journals.

- Use a 70-page spiral notebook or composition book. This helps get rid of the continual "page fallout" that happens in a three-ring binder.

- Handouts, worksheets, tests, or photocopies can be glued, stapled, or taped onto blank pages to keep permanently. The edges of the paper may be trimmed for a neater appearance.

- Page numbering begins with the table of contents (page 1) and continues through the book, on both the fronts and backs of the pages.

Components of a Science Journal

TITLE PAGE. This is designed by the student to reflect the current unit of study. A decision by the teacher is made as to color, font, drawings, and information to include on the front. It is taped or glued to the front of the journal.

GEMS OF WISDOM. This page is glued onto the back of the cover. It is used as a resource for important information, facts, definitions, equations, or diagrams. Students may access this page during an assessment if allowed by the teacher. In some journals, this page may be used as a glossary of terms and phrases.

TABLE OF CONTENTS. The table of contents lists all contents included in the journal. It is the first loose page of the journal and may be several pages long.

STUDENT PAGES. The work completed by students is the most important part of the science journal. Teachers will often use the right side of the journal for the classroom instruction and for information that is given by the teacher. This would include notes, modeling problems, written lab and data sheets, and tests or quizzes. The left side is used by the students to reflect on their work, show their understanding, make connections to the instruction, and share science literacy comprehension.

FEEDBACK PAGE. The final page of the notebook is where teachers provide feedback to the students' parents or guardians about the contents of the journal. The last page of the science journal is designed for parents or guardians to give feedback about the contents of the notebook. Many teachers use this page to provide information regarding the student's progress, successes, and needs. A decision has to be made by the teacher as to whether the journal will be sent home for signatures. In some cases, the journal may be lost or destroyed enroute, so teachers should set high expectations for the use of the journal, in and out of class.

Each teacher has a preference as to how students should keep their science journal. The journal is beneficial not only to students, because teachers are finding additional ways to teach more effectively. Some teachers report that the limited number of pages (seventy) has streamlined the amount of information that is given to students. Many now use the Priority Standards, skills, concepts, and vocabulary to ensure student learning and reflection in the journal. It has served as an excellent source of data, as a way to track student work over time, and as a way to evaluate student progress. Other teachers have used the science journal as a source of information during student-led conferences. No matter how you use the journal, it provides valuable insights into the student's strengths, weaknesses, and challenges with science content.

MASTERING SCIENCE VOCABULARY

Robert Yager's 1983 study on vocabulary suggested that the amount of new vocabulary in science textbooks exceeded the number of vocabulary words for learning a foreign language. More and more, educators are concerned with the number of science terms that must be mastered in order to understand the standards and benchmarks.

Project 2061 began a new reform with state standards. The No Child Left Behind focus on English and math resulted in students entering the middle grades with a deficit and lack of knowledge in their science vocabulary (Cunningham and Allington, 2007).

All teachers must teach vocabulary. Understanding content vocabulary is an excellent predictor of success in the subject area (Wilcox, 2006), and science vocabulary can be overwhelming for many students. Those who have existing problems with reading and comprehension are challenged more with science terminology. Consider the following word: fluorescence. The student may skip the word and become frustrated with the inability to make sense out of the word in the passage. After identifying the key vocabulary by grade level, the word should continue to be included in students' vocabulary at each grade level. As students are exposed to the science terminology each year, the definitions will become more complex.

In teaching science vocabulary, several methods should be considered:

1. Present new vocabulary prior to reading the science text selection.

2. Make resources available and accessible during reading.

3. Incorporate a variety of strategies that reinforce vocabulary acquisition.

4. Use the word in context and in real-world examples.

Identifying Priority Science Standard Vocabulary

Science has an overwhelming amount of vocabulary and complex terminology. The conceptual unit has a place for teachers to identify the priority vocabulary that students should know and be able to use in order to understand the science concept being taught. Grade-level teams should decide on five to seven terms that all students will master during the conceptual unit. Although there will be more words and phrases that are important to know, these are only a few selected terms that represent critical understanding.

Many teachers display these words on a classroom bulletin board or wall. In some cases, the standard is listed in the middle of a large graphic organizer and the words, sometimes with illustrations, are arranged around the center. It is suggested that students create the visual for the display, allowing them to take more ownership of the learning.

For example, a first-grade unit on the water cycle could include these words:

water cycle	evaporation	precipitation	flower
condensation	solid	liquid	gas

Suggested Science Vocabulary Strategies

Students learn new vocabulary when they have opportunities to use multiple modalities to hear, say, read, write, and see the words in a variety of contexts. The goal of vocabulary instruction is to have students use words correctly in scientific speech and in writing communications when explaining science concepts.

The following six strategies will help teachers work with students on internalizing new vocabulary in science.

Mental Mind Maps

A mind map is a diagram used to represent words, ideas, tasks, or other items that are linked to and arranged around a central key word or idea. Mind maps are used to generate, visualize, structure, and classify ideas, and as an aid in study, organization, problem solving, decision making, and writing. Mind maps (or similar concepts) have been used for centuries to improve learning, brainstorming, memory, visual thinking, and problem solving by educators, engineers, psychologists, and others. Some of the earliest examples of mind maps were developed by Porphyry of Tyros, a noted thinker of the third century, as he graphically visualized the concepts of Aristotle. Philosopher Ramon Llull (1235–1315) also used mind maps.

A mental mind map is a learning strategy that helps students to "visualize" vocabulary. The purpose of the strategy is to focus the students' attention, explicitly and effectively, as they "see" the word. For example, if the word is "deciduous," the teacher would have the students look up and to the right to "see" the word. The teacher may ask some visualization questions such as, "How big is this tree?" "Where do you see the tree located?" or "Can you see the leaves?" The objective is to create a strong mental image of the word so that students can recall the word over time.

Vocabulary Webs

This activity uses a graphic organizer that targets a priority vocabulary term that has been identified in the conceptual unit. There are four components to the web: definition, illustration, specific example, and sentence creation. The teacher provides the word to the class and students complete each of the web components. If differentiation is needed, the teacher assigns specific categories to different cooperative groups. For example, one group would do the definition, one would complete an illustration, and so forth. This strategy may be used as a bell or warm-up activity, as a transition, or as homework. A vocabulary web is shown in Exhibit 4.8.

EXHIBIT 4.8: Vocabulary Web

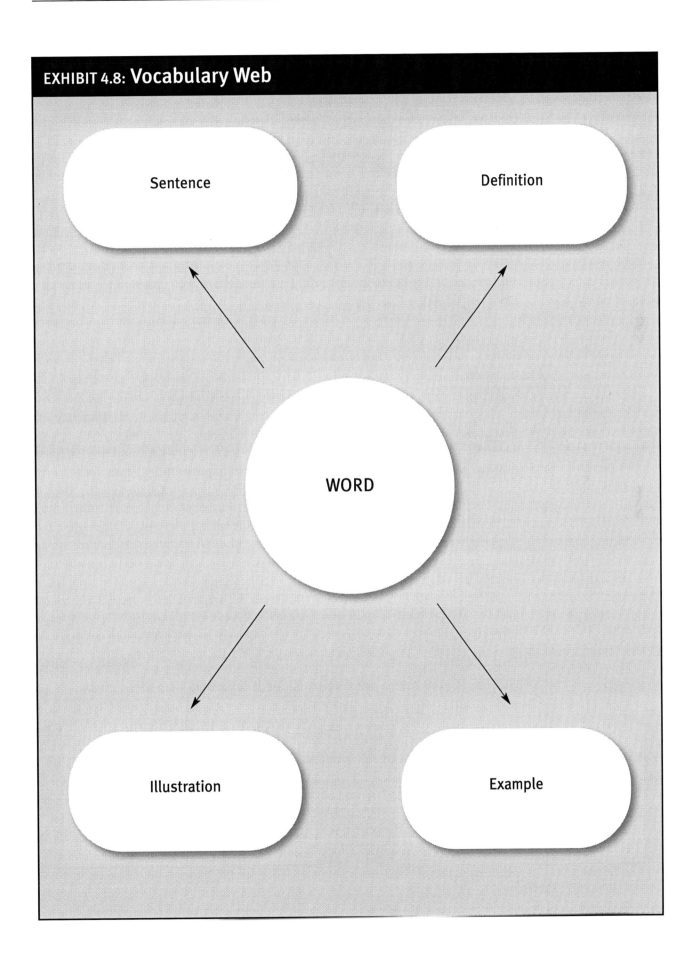

Aerobic Spelling

Spelling should be an integral component of good science instruction. Although many of the science words are difficult and complex, aerobic spelling allows students to have some fun with the vocabulary terms in the conceptual unit. The teacher identifies the priority vocabulary first. Then, as a teacher-directed activity, the teacher says the word for the students. They repeat the word three times, then spell the word with the teacher. However, as the teacher spells the word out loud, students stand up if the letter is a consonant and sit down if the letter is a vowel. Asking students to define the word after spelling holds them accountable for comprehension of the word. This strategy is an excellent transition from one part of the lesson to another.

Vocabulary Chain

The vocabulary chain is an activity that can be used at the beginning or end of class or as a transition between sections of a lesson plan. The teacher identifies the priority vocabulary from the conceptual unit prior to this activity. The purpose is to have students access prior knowledge and build on the knowledge presented in the unit. The teacher begins by stating a vocabulary word. A student is called upon who must give a word that begins with the last letter of the teacher's word.

An example from the astronomy unit might be: A teacher says "solar system," and the first student might say "**Mars**." The next student could say "**Saturn**," and the next response might be "**Nebulae**." The activity continues, with random students being called on around the room. If a student cannot think of a word, he or she may "pass," but the teacher alerts the student that he or she should be ready in a couple of turns. In the lower grades, word walls and resources may help the student be more successful with this activity.

A word wall is a systematically organized collection of words displayed on a wall in the classroom. It is an instructional strategy that teachers use to reinforce key vocabulary in a core subject. Although the word wall is typically used as a primary reading tool, science words, phrases, or visuals can be used to reinforce student understandings about science concepts or vocabulary words. If the word wall is interactive, with students being able to touch or move the words, they will have a fun way to develop stronger relationships with science information.

EXHIBIT 4.9: Template for a Science Loop

Na	O	H	Li
Nitrogen	Carbon	Helium	Beryllium
N	C	He	Be
Oxygen	Hydrogen	Lithium	Sodium

Science Loops

Science loops are one of the most engaging and effective strategies for all grade levels. A template, similar to the one shown in Exhibit 4.9, is created by the teacher and may include vocabulary, definitions, formulas, and other scientific information. For example, a kindergarten teacher may have a picture of an animal that must match up with its habitat. A chemistry teacher may have the name of an element that matches with its chemical symbol. The idea is for the bottom of each column to match with the top of the next column, and the bottom of the last column to match with the top of the first column. If the design is accurate, the cards can be pasted into a circle when they are cut apart.

Vocabulary Brainstorm

The vocabulary brainstorm is a high-level-thinking, compare-and-contrast activity for use with scientific text. Prior to reading a text selection, students are presented with eight to ten words from the text. Half of the words are related directly to the text, and the other half do not have anything to do with the selection. The terms are displayed for the students to discuss which ones belong and which ones do not. After a short discussion with a partner and the class, students read for understanding and clarification of the words.

For example, a first-grade unit on the rainforest might include these eight words:

frog	camel	canopy	iceberg
cactus	camouflage	toucan	sand dune

USING GRAPHIC ORGANIZERS IN SCIENCE

Graphic organizers are visual representations of a student's knowledge, including facts, ideas, and concepts. Using graphic organizers in science allows students to personally interact with the content.

Most include a visual representation, such as a chart, time line, flowchart, or diagram, to record, organize, synthesize, and evaluate information and ideas. Graphic organizers also help to:

1. Relieve learner boredom

2. Enhance recall of content and information

3. Provide motivation

4. Create interest in learning a new topic

5. Clarify information

6. Assist in organizing thoughts

7. Promote understanding of complex concepts

They take many forms:

• **Compare and contrast.** Used to analyze similarities and differences between things (people, places, events, actions, ideas) by listing characteristics on either the left- or right-hand sections. See Exhibit 4.10 for a template.

• **Sequence.** Used to see changes over time, illustrate step-by-step methods, and show complex processes or time lines. See Exhibit 4.11 for a template.

EXHIBIT 4.10: Graphic Organizer: Compare and Contrast

Topic:

EXHIBIT 4.11: Graphic Organizer: Sequence

Topic or Question:

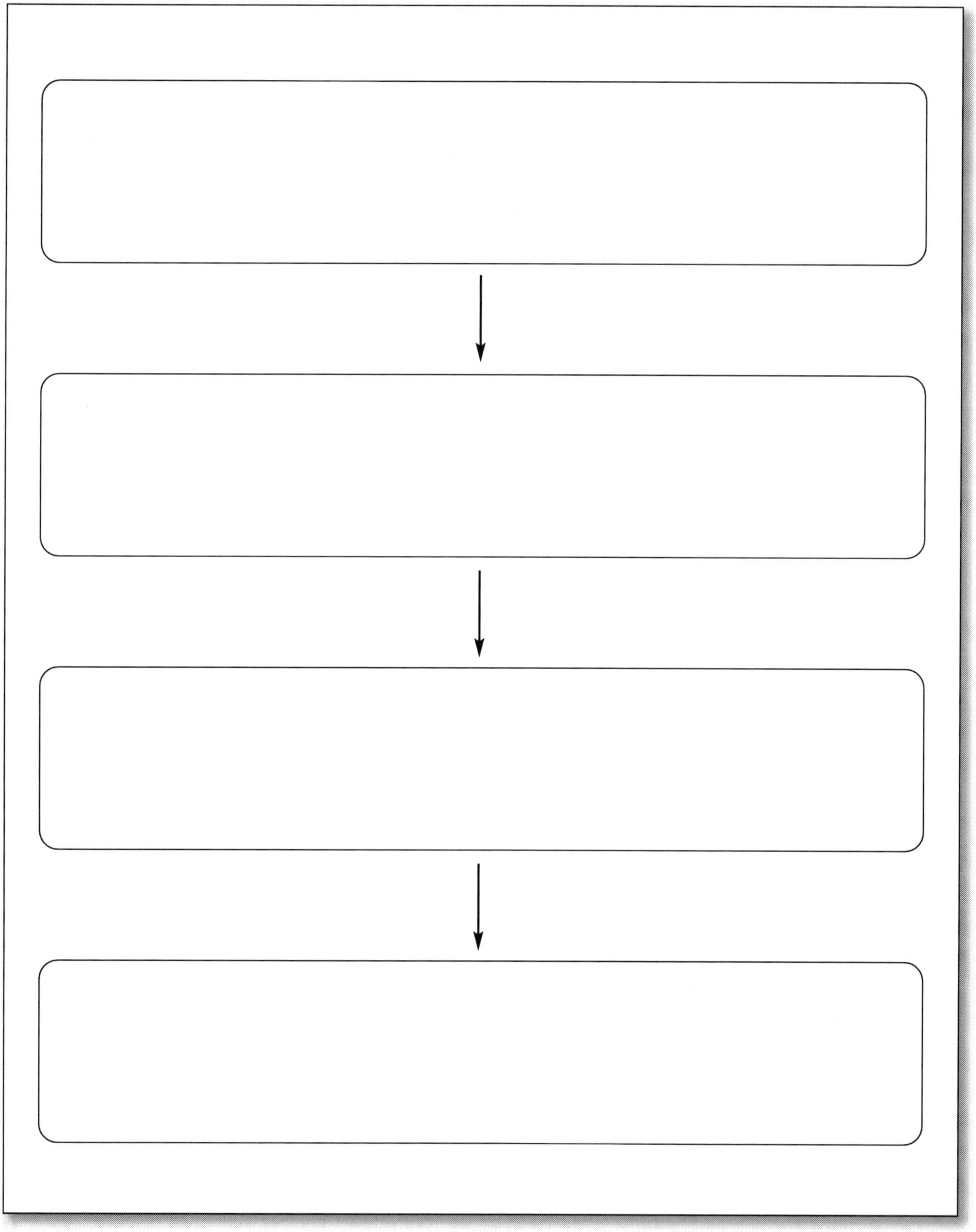

EXHIBIT 4.12: Graphic Organizer: Main Ideas and Details

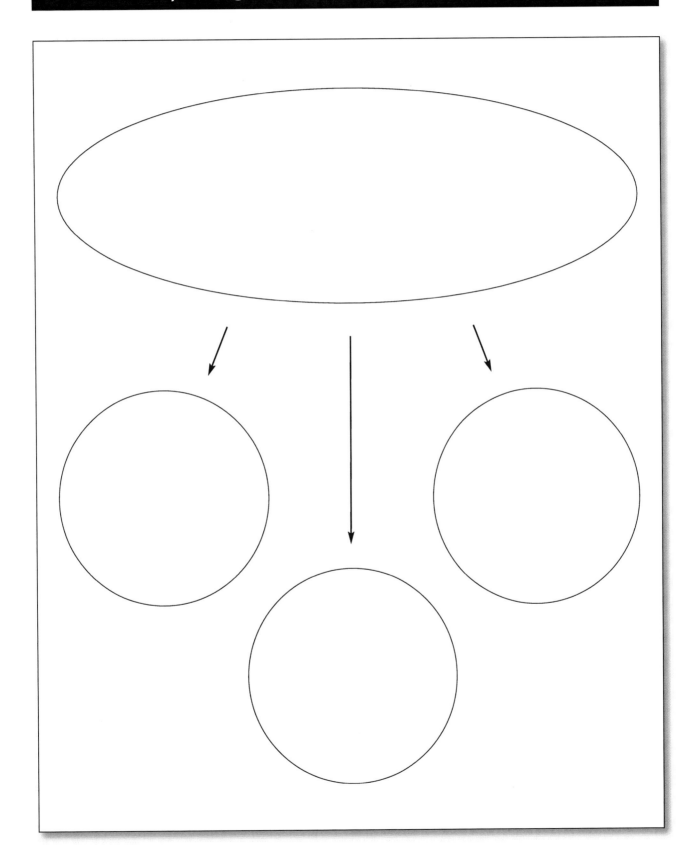

EXHIBIT 4.13: Graphic Organizer: Drawing Conclusions

Topic or Question:

Fact:

Fact:

Fact:

Conclusion:

EXHIBIT 4.14: Graphic Organizer: Text Structures

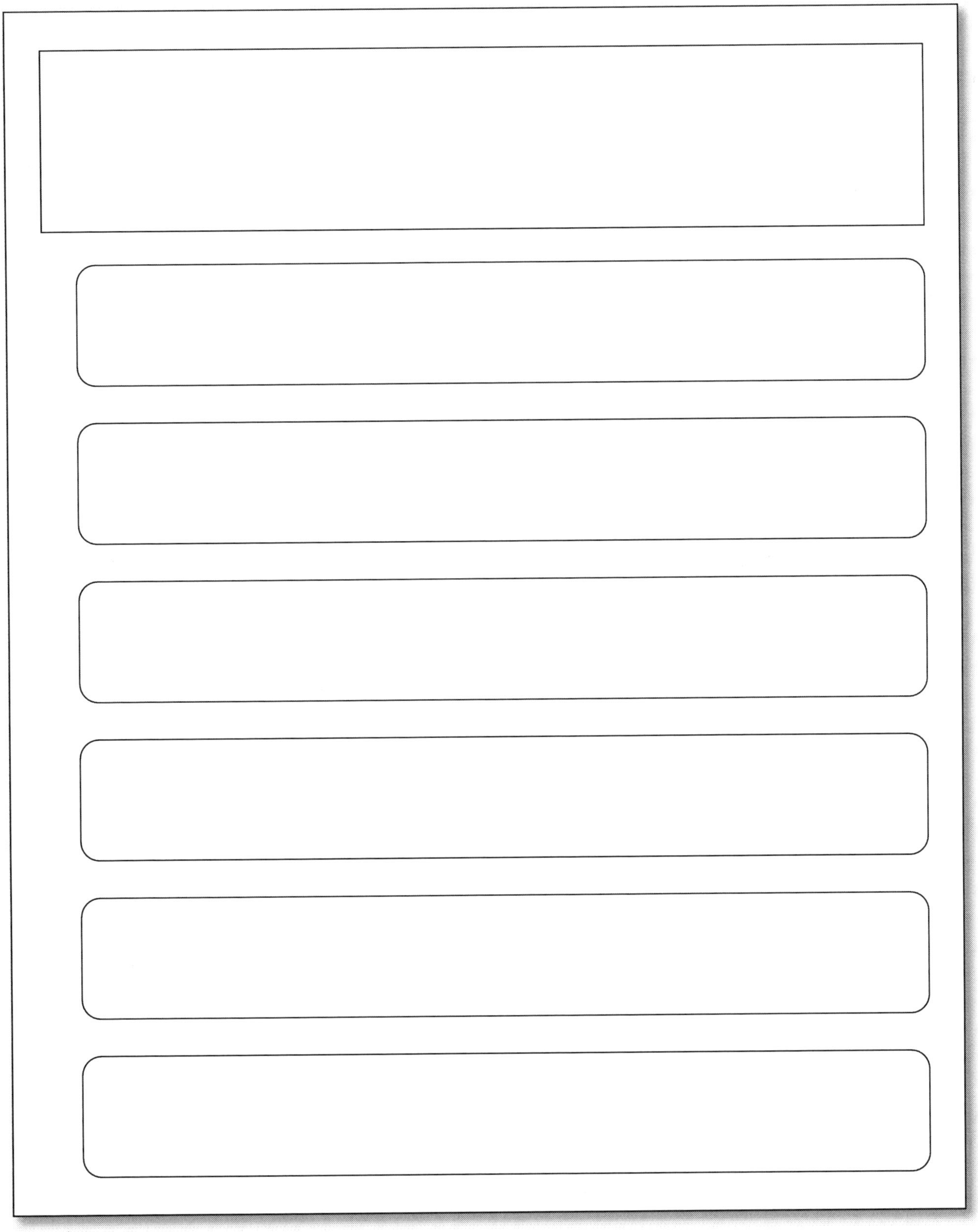

EXHIBIT 4.15: Graphic Organizer: Cause and Effect

Topic or Question:

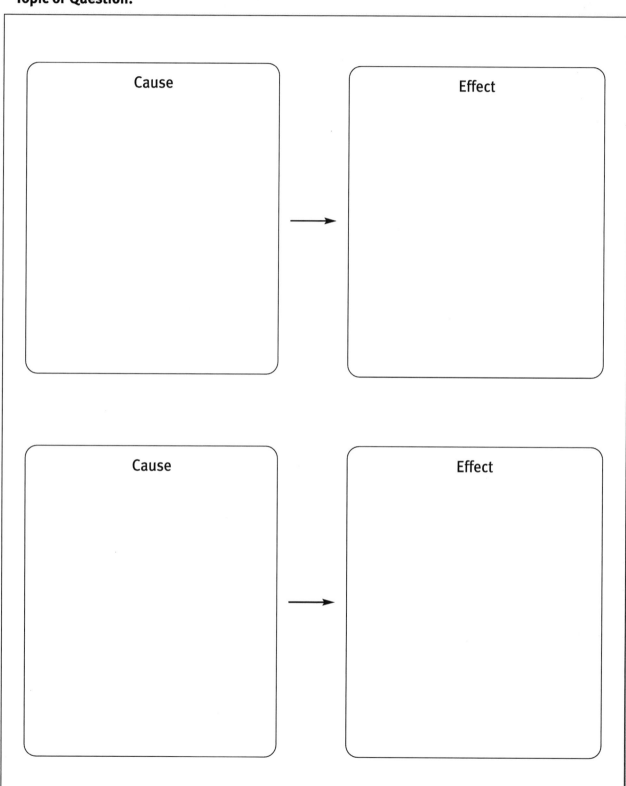

EXHIBIT 4.16: Graphic Organizer: Problem–Solution

Topic or Question:

Problem	Solution
	1.
	2.
	3.

- **Main idea and details.** Used to describe a central idea, process, or concept of a text selection or scientific experiment. See Exhibit 4.12 for a template.

- **Drawing conclusions.** Used to summarize or evaluate an idea, process, or concept of a text selection or a scientific experiment. See Exhibit 4.13 for a template.

- **Text structure.** Used to describe and illustrate the different text structures found in expository text and scientific literature. See Exhibit 4.14 for a template.

- **Cause and effect.** Used to show the relationship between events or actions and the resulting consequences. See Exhibit 4.15 for a template.

- **Problem–Solution.** Used to show the problem-solving process by defining the components of the problem and possible solutions. See Exhibit 4.16 for a template.

Reflecting on My Learning

- What were the most important things to learn in this step?

- What is the value of implementing strategies for mastery?

- How will you incorporate some of the ideas presented in this step?

- What will make the most impact on your students' learning?

STEP 5: Common Formative Assessments

CHAPTER

5

T HE FINAL STEP ALIGNS SCHOOL-BASED ASSESSMENTS *for* learning to science Priority Standards. As Black and Wiliam state in "Assessment and Classroom Learning" (1988), "In reviewing 250 studies from around the world between 1987 and 1988, we found that a focus by teachers on assessment *for* learning, as opposed to assessment *of* learning, produced a substantial increase in students' achievement."

The formative assessments are collaboratively designed, administered, scored, and analyzed within each grade level throughout the school year. Common formative assessments provide teachers with valid feedback regarding the students' current understanding and knowledge of the Priority Ptandards. These data provide value regarding how students are likely to perform on district and state assessments—in time for teachers to modify and adjust instruction to meet specific learning needs.

Essential Question

How are you assessing the effectiveness of science instruction within your grade level, school, or district throughout the school year?

✎ KEY POINTS

1. Regular and timely feedback from common formative assessments allows teachers to modify and adjust instruction to meet the diverse learning needs of all students.

2. Students are allowed to demonstrate scientific understanding through multiple measures of assessments.

RATIONALE BEHIND COMMON FORMATIVE ASSESSMENTS

In *Balanced Assessment: The Key to Accountability and Improved Student Learning* (Cutlip, 2005), the National Education Association (NEA) suggests that assessments can be used "to inform students about the continuous improvements in their achievement and permit them to feel in control of that growth."

HOW POWERFUL PRACTICES WORK TOGETHER

Five Easy Steps to a Balanced Science Program is based on best instructional practices that are designed in a continuing cycle of improvement and reflection. Districts should identify their science Priority Standards before beginning any new initiative. Once the Priority Standards have been reviewed and chosen, the cycle is completed by "unwrapping" the skills and concepts, determining Big Ideas and Essential Questions, and designing appropriate assessments with scoring guides. At the heart of this process lie effective instructional teaching strategies (see Exhibit 5.1). Over time, these pieces are reviewed and revised as needed, but the essential goal is for student achievement and teacher instruction to improve.

EXHIBIT 5.1: Effective Teaching Strategies

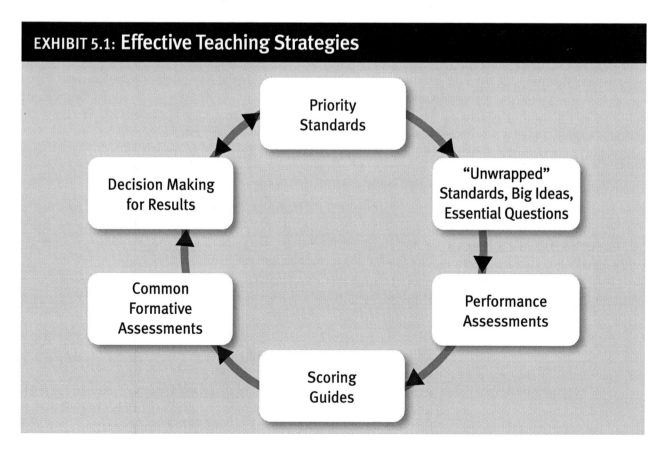

"Schools with the greatest improvements in student achievement consistently used common assessments."

— D.B. REEVES, ACCOUNTABILITY IN ACTION, 2004

WHAT ARE COMMON FORMATIVE ASSESSMENTS?

Common formative assessments are assessments that are collaboratively designed by a grade-level team of teachers and administered to all of the students in that grade level several times during the year. The results of common formative assessments allow teachers to evaluate the level of student understanding of the Priority Standards. The results of the assessments are analyzed through a Data Teams process, with information being shared with all grade-level teachers. These assessments provide timely feedback so that teaching may be differentiated based on the learning needs of their students. The ultimate goal is for the common formative assessment to drive and inform instructional practices.

Common formative assessments:

• Provide a degree of consistency

• Represent common, agreed-upon expectations

• Align with Power Standards

• Help identify effective practices for replication

• Make data collection possible

DATA TEAMS: THE FIVE-STEP PROCESS*

Data Teams are small, grade-level or departmental teams that examine individual student work generated from common formative assessments. These teams attend collaborative, structured, scheduled meetings that focus on the effectiveness of teaching and learning. The Data Teams process includes five steps that are cyclical in design.

Step 1: Collect and chart data and results.

Step 2: Analyze strengths and obstacles.

Step 3: Set goals for student improvement.

Step 4: Select instructional strategies.

Step 5: Determine results indicators.

---

*Source: Data Teams. 2008. The Leadership and Learning Center. Englewood, CO: Lead + Learn Press.

PREPARING THE STANDARDS FOUNDATION

The data provided from summative assessments are often limited and not timely and frequently do not indicate how well students know or have learned the material being taught. The state assessment is an assessment *of* students' learning that is summative, whereas the common formative assessments for classrooms are assessments *for* learning. Both are useful and necessary.

Teachers use pre- and post-assessment results to plan and adjust instruction, lessons, interventions, and informal classroom assessments. The pre-assessment results can be used to identify students who are excelling and those who need extra support. This assessment may guide the differentiation process within the classroom so that teachers can modify assignments within small-group work time. This is especially common in elementary classrooms. Teachers who use the pre-assessment as a "lens" to the post-assessment are providing their students with a format that prepares them for additional testing.

PURPOSE AND BENEFITS OF COMMON FORMATIVE ASSESSMENTS

The purpose of common formative assessments, in general, is to find out what your students know and are able to do with regard to the standards you are teaching. Common formative assessments provide teachers with regular and timely feedback regarding student mastery and understanding of the science Priority Standards. They are designed to be multiple-measure assessments that allow students to demonstrate their scientific understanding in a variety of styles. Each grade level is able to establish consistent expectations for standard implementation, student achievement, instructional practices, and aligned assessments. When assessment results are analyzed in a Data Teams process, there is a predictive value about how students will succeed on additional assessments.

FORMATS AND FOCUS OF COMMON FORMATIVE ASSESSMENTS

Common formative assessments are similar in design and format to end-of-unit assessments and district and state assessments. These assessments should only target the Priority Standards from the conceptual unit. Most common formative assessments contain two types of items: selected response (multiple choice, matching, true/false) and constructed response (short-answer and problem-solving tasks). The common assessment may include items from more than one science standard. Students are required to demonstrate both their procedural understanding (the science process skills) and their conceptual understanding (Big Ideas).

STEPS FOR IMPLEMENTING COMMON FORMATIVE SCIENCE ASSESSMENTS

The following is a suggested sequence of steps to follow when designing and administering a common formative science assessment. These are the steps that were explained in Chapter 3, "Conceptual Understanding," and are listed here only as a reminder of the process.

Design the science conceptual unit:

1. Identify science Priority Standards for each grade level and/or course. This is often done by the district, but individual schools can do this within grade-level teams.

2. Determine important science topics for common formative assessments. The topic of the conceptual unit is found embedded in the standards. This is the main focus for the unit.

3. "Unwrap" the Priority Standards to identify concepts and skills that students need to know and be able to do.

4. Determine Big Ideas and Essential Questions from the "unwrapped" Priority Standards.

After designing the science conceptual unit, continue with the following:

5. Collaboratively design the common pre- and post-assessments that are aligned to the concepts, skills, and Big Ideas from "unwrapped" priority standards, including selected- and constructed-response items.

6. Administer and score the common formative pre-assessment, and analyze the results in grade-level Data Teams.

7. Teach the conceptual unit in each classroom and assess students informally throughout the unit, adjusting instruction as needed.

8. Administer and score the common formative post-assessment, and analyze results in grade-level Data Teams.

9. Design classroom end-of-unit assessments and a scoring guide that are matched to the common formative assessment.

10. Determine the next steps.

IMPORTANT CONSIDERATIONS

No matter what grade level you are teaching, the next section of information is relevant for all elementary and secondary teachers and teams of teachers, whether at the district or school level.

1. Common formative assessments should be aligned with the district benchmarks and end-of-course assessments no matter when they are given. Many districts conduct quarterly assessments as comparative benchmarks from school to school, grade level to grade level, and student to student.

2. Results from the common formative assessments must be analyzed in a Data Team format so that midcourse corrections in teaching and learning can be made before a summative assessment.

3. Common formative assessments must mirror the format of the state and local benchmarks and tests. Released test items are available in most states and serve as a guide to assessment item design.

Exhibit 5.2, adapted from *Common Formative Assessments: How to Connect Standards-Based Instruction and Assessment* (Ainsworth and Viegut, 2006) shows how all of these steps relate to collaboration, design, and implementation of common formative assessments.

GENERAL GUIDELINES FOR WRITING TEST ITEMS

Writing assessment items can be a challenging task for any educator. Several suggestions are included for your consideration in writing assessment items:

- Use clear, straightforward language and terminology that students understand.
- Create bias-free items (e.g., gender, ethnicity, language, religion).
- Answering the questions should not depend on the students' reading or writing ability.
- Match the assessment item to the format in district and state assessments.

Five Roadblocks to Effective Item Writing

If you have designed the common formative assessment to align with the standards, there may still be issues for students taking the assessments. To ensure that students have maximum success on your assessments, consider the following as red-flag indicators:

EXHIBIT 5.2: *Five Easy Steps* Balanced Science Alignment

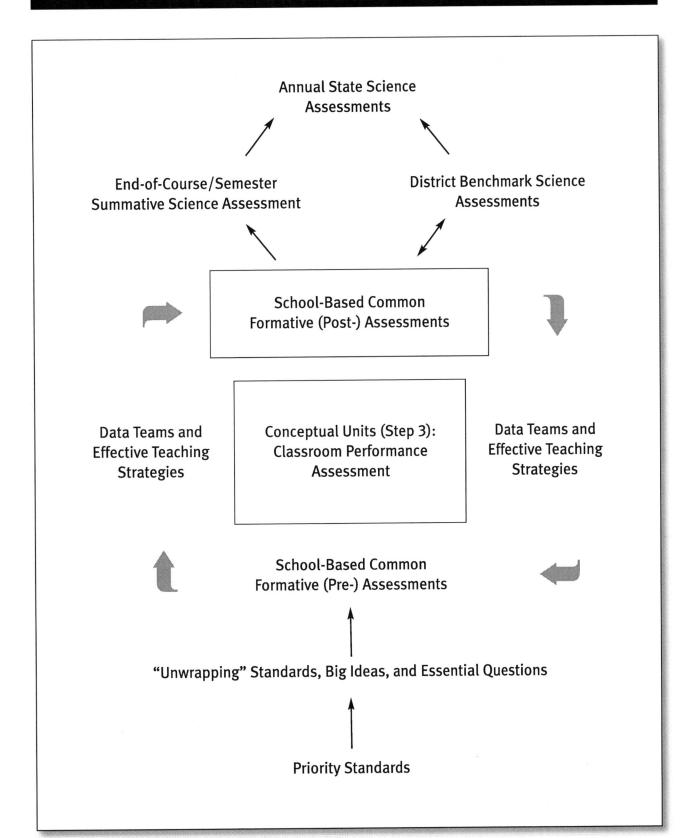

1. Unclear directions in the problem or question

2. Ambiguous statements that are confusing or vague in nature

3. Unintentional clues that are not reasonable distractions

4. Complex phrasing with too many words in the question or answer

5. Difficult vocabulary (not grade-level appropriate) in both the problem and the answer

How Many Items to Write?

There is always a question from grade-level teams about how many items should be included in a common formative assessment. By using "multiple-measure" assessments, educators are better able to make more accurate inferences about student learning and understanding. It is also important that teacher instructional time focus on the Priority Standard as it is matched to the common formative assessment.

It is suggested that the assessment include:

• Four to six selected-response items (consider that most state tests are designed in the multiple-choice format with no true/false, matching, or fill-in-the-blank items)

• One constructed-response item with an accompanying scoring guide

• Sufficient representation of each Priority Standard

• Sufficient representation of the standards being measured on the state test (e.g., one objective that is known to be heavily tested on the state test should have equal weight in the common formative assessment)

TYPES OF RESPONSE ITEMS FOR COMMON FORMATIVE ASSESSMENTS

There are three kinds of response items for common formative assessments, as depicted in Exhibit 5.3.

Selected-Response Items

Selected-response items require students to select one response from a list of answers. Types of selected responses include multiple-choice, true/false, matching, and fill-in-the-blank. Most state tests are designed in the multiple-choice format, so it is suggested that grade-level teams use this configuration when creating selected-response items for their common formative assessments.

EXHIBIT 5.3: Response Items for Common Formative Assessments

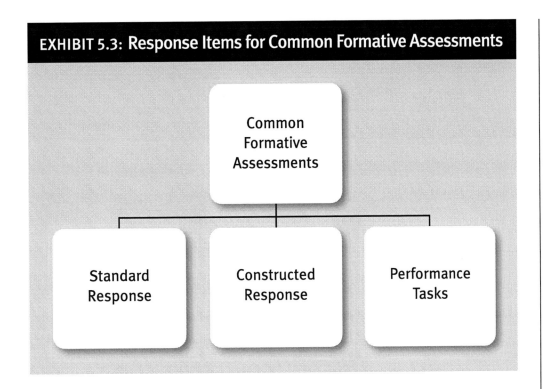

One of the benefits of selected-response items is that they can be answered quickly by the students and can be objectively scored as correct or incorrect. This effectively assesses students' knowledge of factual information, priority vocabulary, and basic skills. However, this type of question tends to promote recall or memorization of facts rather than demonstrate evidence of higher-level thinking and understanding. Selected-response items do not support writing and creative thinking unless they are specifically designed to do this.

Criteria for Writing Selected-Response Items

1. Write clearly in a sharply focused manner.

2. Ask a question with only one best answer.

3. Write items that are consistent with grade-level reading expectations.

4. Eliminate clues leading to the correct answer.

5. Make response options brief.

Constructed-Response Items

This is the portion of the common formative assessment where students will demonstrate through writing, speaking, or performance their integrated understanding of the "unwrapped" concepts and skills—at the level of rigor specified in the standard. For example, if the standards say "describe," then the constructed-response item would reflect this cognitive skill.

Constructed-response items include short responses (word, phrases, and sentences, problem-solving steps, or multiple sentences or paragraphs). This type of question requires students to organize their thinking and use concept knowledge and skills or to answer a question or complete a task. Teachers are able to see whether or not students have gained an understanding of the Priority Standard. The benefit of constructed-response questions is that they provide the teacher with a valid inference about student knowledge. However, these take longer to score and require a scoring guide to determine the levels of proficiency.

Common formative assessments can be written so that students *physically* demonstrate their understanding of the targeted concepts and skills. Their performance and corresponding degree of proficiency can then be evaluated by using a scoring guide, or rubric. Exhibit 5.4 provides a template for designing a common formative assessment.

Performance Tasks

Performance tasks give students the opportunity to demonstrate the level to which they have mastered the Priority Standard. These open-ended products and performances are designed to show what the student should know and be able to do after completing a conceptual unit. Performance tasks should consist of complex challenges that reflect current science trends.

Science performance tasks should:

- Allow students to demonstrate and communicate their knowledge of standards, goals, objectives, and science content
- Allow students to demonstrate their understanding of science by requiring them to demonstrate more than one plausible solution
- Integrate science with other core content disciplines
- Provide meaning, interest, challenge, and real-world relevance to the content
- Allow for a deep investigation into the task

The Office of Technology Assessment of the U.S. Congress defines performance assessment as "any form of testing that requires a student to create an answer or a product that demonstrates his or her knowledge or skills."

EXHIBIT 5.4: Designing a Common Formative Assessment

Standard(s)			
Topic			
"Unwrapped" Skills		**Content or Concepts**	
Science Process Skills			
Big Ideas		**Essential Question(s)**	
Selected-Response Items		**Constructed-Response Items**	
Performance Task 1.	2.	3.	4.

Effective performance assessments contain tasks that are engaging and worthwhile for the student. Performance assessments contribute to student learning by giving them challenging, high-level tasks that require them to apply skills and knowledge learned prior to the assessment. Many teachers find that performance tasks developed individually or by grade-level teams provide better data about student learning than traditional assessment methods.

Designing a Performance Task

Good performance tasks require that students use higher-level thinking and the science process skills in solving the question or task, and they are more aligned with the National Science Education Standards.

For example, K–2 students were asked to complete a performance assessment after completing a unit on the Living Environment. The task involved having students observe, collect information, measure, and classify objects in the natural environment. This could include leaves, rocks, or sand samples. Students were asked to review their content knowledge, then conduct an investigative walk around the outdoor environment. The tasks involved simple observations, collecting scientific samples, organizing data (compare and contrast), using appropriate tools, and communicating their results. In addition, teachers could have students track change over time with a focus on the seasons as leaves change during the year.

The following guidelines may be followed as teachers develop performance tasks based on a conceptual unit in science.*

- Ask the questions "What are we going to do in this task? What do we want our students to accomplish?"
- Develop an overview of the performance assessment, including:
  - An engaging scenario
  - Three to four tasks to be completed
  - Student understanding of the Big Ideas
  - Differentiation for varying abilities
  - Nonfiction and fiction connections to the task
  - Multiple levels of thinking skills

---

*Source: Adapted from The Leadership and Learning Center's *Engaging Classroom Assessments* seminar.

– The time frame for completion

– A scoring guide for evaluation

• Develop student understanding of the Big Ideas.

• Differentiate for varying abilities.

• Include nonfiction writing in at least one task.

• Scaffold (increase the complexity of) the task to build understanding of concepts and skills.

• Determine the materials needed.

• Establish a time frame for completion.

Developing the Performance Tasks

Use the template in Exhibit 5.5 as a guide for designing each performance task.

As teachers design performance tasks, they should consider these four areas: the science process skills, cognitive thinking levels, the products to be completed, and the type of presentation modality. Use Exhibit 5.6 to help guide your thinking about each task.

SCORING GUIDE

A scoring guide or rubric is a set of criteria used to evaluate student performance on a specific task or constructed-response item. The purpose is to describe "proficiency" in specific terms that a student must meet in order to demonstrate knowledge of a Priority Standard or set of standards. Scoring guides should be shared with students prior to their beginning work. Students are provided timely feedback on their work and may be given another chance to move to a proficient level.

The following are characteristics of an effective scoring guide for formative assessments:

• Establishes a set of general and/or specific criteria used to evaluate student performance on a constructed-response task

• Describes "proficiency," in specific terms, as the level of performance students must meet to demonstrate attainment of particular standards upon which a performance task is based

• Identifies the degree or level of proficiency the student achieves at the time of scoring

EXHIBIT 5.5: Template for Designing Performance Tasks

Which Priority Standard(s) does the performance task address?
Which "unwrapped" skills and concepts are targeted in this task?
What are the corresponding Big Ideas and Essential Questions for the performance task?
Which science process skills are to be demonstrated during the task?
What cognitive skills will be included in the task?
What resources, materials, equipment, and/or supplies does the task require?
What evidence of learning will the student demonstrate for conceptual understanding of the standard? Students score a "proficient" level or higher. Students respond to Essential Question(s) with Big Idea(s) in written or oral format. Other evidence:
What differentiation strategies must be included to meet the learning needs of the students?

EXHIBIT 5.6: Guide to Thinking about Performance Tasks

Science Process Skills	Cognitive Thinking Verbs	Products	Modalities
Observing	Knowledge	Narrative	Visual
Communicating	Comprehension	Expository	Linguistic/Written
Classifying	Application	Descriptive	Kinesthetic
Measuring	Analysis	Persuasive	Verbal/Oral
Inferring	Synthesis		
Predicting	Evaluation		

- Contains specific language understood by all: students, teachers, and parents
- Is available to be referred to frequently during the completion of the task
- Is used to assess the completed task
- Expedites the evaluation of student work and helps provide timely feedback on student performance

Begin with "Proficient"

The goal for all students is to demonstrate proficiency or beyond on the constructed response. As teachers design scoring guides, it is best to start with a decision on what is "proficient" for the specific task or question. These are often referred to as "nonnegotiables" for showing that learning has occurred. Once the proficient criteria have been established, teachers then determine "exemplary," "progressing," and "beginning" levels of performance. A template for a constructed-response scoring guide is presented in Exhibit 5.7.

EXHIBIT 5.7: Template for a Constructed-Response Scoring Guide

Name:

Exemplary:

Meets all "Proficient" criteria PLUS:

Proficient:

Progressing:

Meets three of the "Proficient" criteria

Beginning:

Meets fewer than three of the "Proficient" criteria

Task to be repeated after remediation

HOW DO TEACHERS USE ASSESSMENT RESULTS?

• Teams of teachers analyze assessment results to diagnose student learning needs.

• Timely and regular feedback allows teachers to differentiate and adjust instruction appropriately.

• The Data Teams process provides a structure for the analysis and review of common formative assessments.

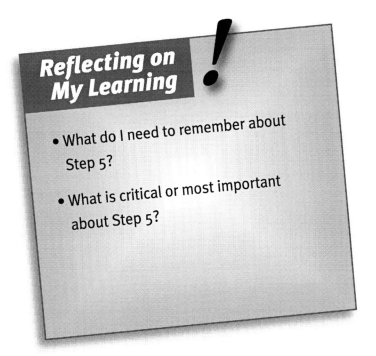

Reflecting on My Learning

• What do I need to remember about Step 5?

• What is critical or most important about Step 5?

PART TWO: Inside the Classroom

I N THIS PART, the application of the five steps is presented by grade levels. The examples provided illustrate problem solving, a sample conceptual unit, the weekly science review, and science literacy strategies. These grade-level examples are designed to benefit teachers in implementing a balanced science program.

Inside the Kindergarten Classroom

CHAPTER 6

STEP 1: ESTABLISHING AN EFFECTIVE SCIENCE ENVIRONMENT

ESTABLISHING AND MAINTAINING AN EFFECTIVE LEARNING ENVIRONMENT is not something you learn from a textbook, science education course, or training manual. It is the expertise and knowledge that you gain over time regarding what really works. It includes your personal teaching style, your students' needs and learning modalities, and having time to practice different techniques and designs.

An early elementary classroom should be inviting and student centered. These classrooms should be developmentally appropriate so that student self-esteem, diversity, collaboration, and independence are established through the various activities. The curriculum should teach the foundations of science through the science process skills. Kindergarten through second-grade classrooms should focus on the following:

- Teachers clearly define the safety rules and procedures for using science equipment and materials. Remember, safety is the number one priority.

- A structure is in place for "playing" and working with materials and other students.

- Students have access to a variety of science materials (e.g., centers, tubs of items, nonfiction and fiction science literature).

- Large- and small-group instruction is integrated throughout the day.
- Student work, with the science content, is displayed and changed with each unit focus.
- Students have an opportunity to read, write, and "do" science.
- The teacher reads aloud from science trade books or relevant literature.
- Students have an opportunity to share their experiences with science and the world around them.
- Time is provided for students to practice the science process skills (e.g., observing, measuring, inferring, communicating, predicting, and classifying).
- Students are allowed to experiment and report their findings.
- The teacher uses appropriate science vocabulary during instruction and checks for student understanding.
- There is a direct connection of life, earth, and physical science to the students' world and life outside of the classroom.

STEP 2: PROBLEM SOLVING

As teachers introduce the problem-solving process to students, keep in mind that consistency throughout the year will make it easier for students to complete the problems. As the year progresses, the teacher should move from teacher-directed modeling to a whole class to doing the problem-solving task together. To keep students accountable, early elementary teachers may have to record the student responses on the board or chart paper.

As the year progresses, the students will record their thinking before the teacher has a class discussion. The goal is to have students make their own data sheet and write the solution on their own.

A suggested instructional sequence is given below.

1. The teacher introduces the selected problem to the students.
2. Students, working alone or with a partner, work on a solution to the problem using writing or graphic interpretations to represent the solution.
3. Students are asked to share their ideas about how they solved the problem. This should include the related content and science process skills.
4. The teacher records some of the ideas on chart paper (this is the class data sheet).

5. The teacher has a quick discussion with the class to determine the best solution.

6. Students are asked to write several sentences or to use illustrations, on their own, which describe the solution to the problem.

Problem-Solving Task

This particular problem is aligned to the kindergarten grade standard of Weather and is aligned to the conceptual unit described in Step 3.

> **Y**ou are the classroom meteorologist and are responsible for reporting the weather to your friends. Take a look at the weather oustide or use one of the pictures provided by your teacher, and be ready to give a detailed weather report to the class. Remember to include the types of clouds, the temperature, wind, and precipitation. You may use a drawing in your report.

STEP 3: CONCEPTUAL UNDERSTANDING UNIT FOR KINDERGARTEN

Exhibit 6.1 is a sample Conceptual Understanding Unit for kindergarten matched to the standard of Weather. It is aligned with the problem-solving task, the weekly science review, and the sample science literacy strategies.

STEP 4: MASTERY OF SCIENCE INFORMATION

Weekly Science Reviews

The science review occurs daily at the beginning of the science block of instruction. Students should complete the problems independently using paper and pencil while the teacher makes modifications to ensure that all students are successful working on their own. For early grades, teachers may want to give one or two of the "blocks" per day rather than all six.

The problems can be written on the board, prepared in advance as an individual worksheet for each student, or given verbally as a start to the instructional day.

There are three types of weekly science review templates. An example of a content-based template (Exhibit 6.2) is provided as related to this kindergarten conceptual unit. The questions are designed to target grade-level standards to

EXHIBIT 6.1: Example of a Conceptual Understanding Unit for Kindergarten

Grade Level: Kindergarten   **Conceptual Unit Focus: Weather**

Standards and Indicators Matched to Unit Focus:

Teachers will list and "unwrap" the full text of their standards and indicators from individual state or district documents. *Source*: North Carolina Public Schools (www.ncpublicschools.org), North Carolina Standard Course of Study for Science, Competency Goal 2 (2.01, 2.02, 2.03, 2.04, 2.05)

Need to Know About Weather Concepts:

❏ Daily Weather Changes   ❏ Types of precipitation   ❏ Effects of weather (on human activities)

❏ Weather features   ❏ Change in   ❏ Common tools (to measure weather)

- Precipitation   - Wind
- Wind   - Force
- Temperature   - Direction
- Cloud cover   - Sky conditions

"Unwrapped" Skills: Be Able to Do:

❏ Observe   ❏ Determine

❏ Report   ❏ Use

❏ Identify

Science Process Skills:

❏ Observing   ❏ Measuring

❏ Communicating   ❏ Inferring

❏ Classifying   ❏ Predicting

Priority Vocabulary:

❏ Weather   ❏ Temperature

❏ Precipitation   ❏ Clouds and cloud cover

❏ Wind   ❏ Thermometer

EXHIBIT 6.1: Example of a Conceptual Understanding Unit for Kindergarten

Grade Level: Kindergarten   **Conceptual Unit Focus: Weather**

Big ideas:

1. Weather happens all the time.

2. Weather changes over time and can be observed, measured, and described.

3. Meteorologists can predict the weather using scientific tools.

Essential Questions:

1. How do we know weather is all around us?

2. How can we observe changes in the weather?

3. What tools do meteorologists use to predict the weather?

Materials:

To be determined by the teacher

Writing Prompt(s):

• How do scientists predict the weather?

• What makes the weather?

• How can we tell if the weather is going to change?

End-of-Unit Assessment

Each state has instituted ways to hold school districts and schools accountable for teaching the standards to the students. Both formative and summative assessments are used to determine proficiency in core content subject areas. The end-of-unit assessment may include a variety of test items, including selected response, constructed response, and performance tasks. End-of-unit assessments in science must include the problem-solving step in conjunction with the science process skills. For this reason, *Five Easy Steps* recommends that grade-level teams of teachers design and implement end-of-unit assessments based on their curriculum standards, the needs of their students, and state and local mandates.

EXHIBIT 6.2: Examples of Science Reviews for Kindergarten

Plants and Animals	Weather	The Senses
Life Science	**Earth Science**	**Physical Science**
Scientific Tools and Measurement	Observing	Vocabulary Aligned with Conceptual Unit
Measurement	**Science Process Skill**	**Vocabulary**

which students are exposed throughout the year. Primary teachers have the discretion to use fewer boxes and to incorporate visual, auditory, and kinesthetic methodologies appropriate to the grade level.

For the examples in Exhibit 6.2, you will have to insert the specific science strands for your individual district or school. The boxes contain topics that fall under each of the categories represented by the strands. It is suggested that you find or develop questions or problems related to your state objectives and benchmarks represented by each strand.

Using Graphic Organizers

This unit has an unlimited possibility of ideas for using graphic organizers. Exhibit 6.3 provides samples of graphic organizers that can be incorporated into the Weather lessons and activities. Teachers should consider age-appropriate use of any of these in their classrooms.

Mastering Kindergarten Science Vocabulary

The early grades require that students learn science vocabulary related to the standards. Grade-level teams of teachers should identify the priority

EXHIBIT 6.3: Examples of Graphic Organizers

■ **Compare and Contrast**

A T-chart would show the difference between the seasons and daily weather patterns.

■ **Sequence**

Day-to-day observations of temperature, cloud cover, and precipitation would be illustrated in a graphic organizer for sequence.

■ **Main Idea and Details**

Stories about weather and season would be read aloud to the students. After reading, the students could write or illustrate the main idea from the story. Make sure that they give supporting details.

■ **Draw Conclusions**

An experiment with temperature or precipitation would allow students to draw conclusions about specific weather phenomena.

■ **Text Structure**

An expository story selection about animals or plants would show students the major text structures found in nonfiction literature. Students would identify the main text structures that are necessary for reading this type of text.

■ **Cause and Effect**

This graphic organizer would be used during a problem-solving activity. For example, "If you see a cumulus cloud in the sky, what do you think the weather will be?"

■ **Problem–Solution**

Students would be presented with a real-world problem and asked to provide a solution. Supporting details should be included with the solution.

vocabulary related to the standards in the conceptual unit. Students should be presented with new vocabulary through visual, written, oral, and kinesthetic activities. The science journal is a place where students can record their own interpretations of the vocabulary during the unit.

Chapter 4 provides several strategies for supporting vocabulary acquisition in kindergarten. Many of these activities could be recorded in the student's science journal.

STEP 5: COMMON FORMATIVE ASSESSMENTS

Common formative assessments for the conceptual unit should be designed by grade-level teams of teachers. It is important to remember that the common formative assessment be aligned only with the Priority Standards taught in the conceptual unit. Because there are so many variations in state standards and primary grade-level assessments, there is not a common formative assessment sample for this unit.

Inside the First-Grade Classroom

CHAPTER 7

STEP 1: ESTABLISHING AN EFFECTIVE SCIENCE ENVIRONMENT

ESTABLISHING AND MAINTAINING AN EFFECTIVE LEARNING ENVIRONMENT is not something you learn from a textbook, science education course, or training manual. It is knowledge of what works for you and an expertise that develop over time. It includes your personal teaching style, your students' needs and learning modalities, and having time to practice with different techniques and designs.

An early elementary classroom should be inviting and student centered. These classrooms should be developmentally appropriate so that student self-esteem, diversity, collaboration, and independence are established through the various activities. The curriculum should teach the foundations of science through the science process skills. Kindergarten through second-grade classrooms should focus on the following:

- A structure is in place for "playing" and working with materials and other students.

- Teachers clearly define the safety rules and procedures for using science equipment and materials.

- Students have access to a variety of science materials (e.g., centers, tubs of items, nonfiction and fiction science literature).

- Large- and small-group instruction is integrated throughout the day.
- Student work, with the science content, is displayed and changed with each unit focus.
- Students have an opportunity to read, write, and "do" science.
- The teacher reads aloud from science trade books or relevant literature.
- Students have an opportunity to share their experiences with science and the world around them.
- Time is provided for students to practice the science process skills (e.g., observing, measuring, inferring, communicating, predicting, and classifying).
- Students are allowed to experiment and report their findings.
- The teacher uses appropriate science vocabulary during instruction and checks for student understanding.
- There is a direct connection of life, earth, and physical science to the students' world and life outside of the classroom.

STEP 2: PROBLEM SOLVING

As teachers introduce the problem-solving process to students, keep in mind that consistency throughout the year will make it easier for students to complete the problems. A suggested instructional sequence is given below.

1. The teacher introduces the selected problem to the students.

2. Students, working alone or with a partner, work on a solution to the problem using writing or graphic interpretations to represent the solution.

3. Students are asked to share their ideas about how they solved the problem. This should include the related content and science process skills.

4. The teacher records some of the ideas on chart paper (this is the class data sheet).

5. The teacher has a quick discussion with the class to determine the best solution.

6. Students are asked to write several sentences, on their own, which describe the solution to the problem.

As the year progresses, the teacher should move from teacher-directed modeling to a whole class to doing the problem-solving task together. At this point, the students will record their thinking before the teacher has a class discussion. The goal is to have students make their own data sheet and write the solution on their own.

Problem-Solving Task

This particular problem is aligned to the first-grade standard of Structure and Function and is aligned to the conceptual unit described in Step 3.

You are a member of a scientific research team. A box has just arrived at your site and you open it to find an unidentified animal skull, miscellaneous bones, and other artifacts, including fur, teeth, and claws. Your job is to hypothesize about the type of animal that it came from. What did the animal use the other pieces for in order to survive?

Below is a description of each article:

The skull is six inches long.

The teeth are two inches long and are very sharp.

The fur is very black, very thick, and coarse.

There are five claws in the box, and each is very sharp.

STEP 3: CONCEPTUAL UNDERSTANDING UNIT FOR FIRST GRADE

Exhibit 7.1 is a sample Conceptual Understanding Unit for first grade matched to the standard of Structure and Function. It is aligned with the problem-solving task, the weekly science review, and the sample science literacy strategies.

STEP 4: MASTERY OF SCIENCE INFORMATION

Science Reviews

The science review occurs daily at the beginning of the science block of instruction. Students should complete the problems independently using paper and pencil while the teacher makes modifications to ensure that all students are successful working on their own. For early grades, teachers may want to give one or two of the "blocks" per day rather than all six.

The problems can be written on the board, prepared in advance as an individual worksheet for each student, or given verbally as a start to the instructional day.

EXHIBIT 7.1: Example of a Conceptual Understanding Unit for First Grade

Grade Level: First Grade   **Conceptual Unit Focus: Structure and Function**

Standards and Indicators Matched to Unit Focus:

Teachers will list and "unwrap" the full text of their standards and indicators from individual state or district documents. *Source*: Connecticut Department of Education (http://www.ct.gov/sde), Connecticut Prekinder-garten – Grade 8 Science Curriculum Standards Including Grade-Level Expectations 1.2 (1.2.1, 1.2.2, 1.2.3, 1.2.4)

"Unwrapped" Concepts: Need to Know About Structure and Function:

❑ Animals and plants (need water, food, and air)

❑ Animal behavior (based on their physical features)

❑ Animal and plant structures

❑ Scientists use tools to measure the effects of water and sunlight on plant growth

❑ Animals and plants (in literature)

"Unwrapped" Skills: Be Able to Do:

❑ Infer   ❑ Use

❑ Identify   ❑ Compare

❑ Sort   ❑ Contrast

❑ Classify

Topic or Context (specific lessons, textbook pages, learning activities used during the unit):

❑ Science textbook

❑ Other resources as determined by the teacher

Science Process Skills:

❑ Observing   ❑ Inferring

❑ Communicating   ❑ Predicting

❑ Classifying

Priority Vocabulary:

❑ Structure of Animals   ❑ Fins   ❑ Claws

❑ Legs   ❑ Gills   ❑ Fingers

❑ Wings   ❑ Lungs

EXHIBIT 7.1: Example of a Conceptual Understanding Unit for First Grade

Grade Level: First Grade   **Conceptual Unit Focus: Structure and Function**

Big ideas:

1. Living things have different structures and behaviors that allow them to meet their basic needs.

2. Animals and plants need air, water, and food to survive.

3. The effects of water and sunlight can be observed and measured.

4. Animals and plants represented in literature often have unrealistic characteristics.

Essential Questions:

1. What do living things need to survive?

2. How does an organism's structure affect its ability to survive?

3. How can we measure the effect of sunlight and water on living things?

4. What is the difference in animals found in fiction and nonfiction literature?

Materials Needed:

To be determined by the teacher

Writing Prompt(s):

• If you were an animal, what would you need to live?

• What do you think helps animals and plants survive?

• Sometimes stories about plants and animals are not true. Write a story about how the zebra got its stripes.

• What would happen if your plant wasn't able to get sunlight?

• Design an experiment for a kindergarten class that shows how plants are affected by sunlight and water.

• Can you think of a story where the animal is not real? Write your own story about an imaginary animal.

End-of-Unit Assessment

Each state has instituted ways to hold school districts and schools accountable for teaching the standards to the students. Both formative and summative assessments are used to determine proficiency in core content subject areas. The end-of-unit assessment may include a variety of test items, including selected response, constructed response, and performance tasks. End-of-unit assessments in science must include the problem-solving step in conjunction with the science process skills. For this reason, *Five Easy Steps* recommends that grade-level teams of teachers design and implement end-of-unit assessments based on their curriculum standards, the needs of their students, and state and local mandates.

EXHIBIT 7.2: Examples of Science Reviews for First Grade

Needs of Plants and Animals	Properties of Rocks and Soil	Liquids and Solids
Life Science	**Earth Science**	**Physical Science**
Balancing, Weighing, and Motion of Objects	Communicating	Vocabulary Aligned with Conceptual Unit
Measurement	**Science Process Skill**	**Vocabulary**

There are three types of weekly science review templates. An example of a standards-based template (Exhibit 7.2) is provided as related to this first-grade conceptual unit. The questions are designed to match the core content standards to which students are exposed during the year. Teachers do have the discretion to use fewer boxes based on student needs.

For the examples in Exhibit 7.2, you will have to insert the specific science strands for your individual district or school. The boxes contain topics that fall under each of the categories represented by the strands. It is suggested that you find or develop questions or problems related to your state objectives and benchmarks represented by each strand.

Using Graphic Organizers

This unit has an unlimited possibility of ideas for using graphic organizers. Exhibit 7.3 provides samples of graphic organizers that can be incorporated into the Structure and Function lessons and activities. Teachers should consider age-appropriate use of any of these in their classrooms.

EXHIBIT 7.3: Examples of Graphic Organizers

■ **Compare and Contrast**

Students could compare and contrast the structures of different animals or plants.

■ **Sequence**

Students could illustrate the process steps for showing that plants need food, water, air, and sunlight.

■ **Main Idea and Details**

After reading a story about an animal or plant, have students write or illustrate the main idea of the story. Make sure that they provide supporting details.

■ **Draw Conclusions**

A fiction story about an imaginary animal could help students draw conclusions about the real or nonrealistic world.

■ **Text Structure**

An expository story selection about animals or plants would show students the major text structures found in nonfiction literature. Students would identify the main text structures that are necessary for reading this type of text.

■ **Cause and Effect**

This graphic organizer would be used during a problem-solving activity. For example, "If a plant does not receive enough light, what would happen?"

■ **Problem–Solution**

Students would be presented with a real-world problem and asked to provide a solution. Supporting details should be included with the solution.

Mastering First-Grade Science Vocabulary

The early grades require that students learn science vocabulary related to the standards. Grade-level teams of teachers should identify the priority vocabulary related to the standards in the conceptual unit. Students should be presented with new vocabulary through visual, written, oral, and kinesthetic activities. The science journal is a place where students can record their own interpretations of the vocabulary during the unit.

Chapter 4 provides several strategies for supporting vocabulary acquisition in first grade. Many of these activities could be recorded in the students' science journals.

STEP 5: COMMON FORMATIVE ASSESSMENTS

Common formative assessments for the conceptual unit should be designed by grade-level teams of teachers. It is important to remember that the common formative assessment be aligned only with the Priority Standards taught in the conceptual unit. Because there are so many variations in state standards and primary grade-level assessments, there is not a common formative assessment sample for this unit.

Inside the
Second-Grade
Classroom

CHAPTER

STEP 1: ESTABLISHING AN EFFECTIVE
SCIENCE ENVIRONMENT

ESTABLISHING AND MAINTAINING AN EFFECTIVE LEARNING ENVIRONMENT is not something you learn from a textbook, science education course, or training manual. It is knowledge of what works for you and an expertise that develop over time. It includes your personal teaching style, your students' needs and learning modalities, and having time to practice with different techniques and designs.

An early elementary classroom should be inviting and student centered. These classrooms should be developmentally appropriate so that student self-esteem, diversity, collaboration, and independence are established through the various activities. The curriculum should teach the foundations of science through the science process skills. Kindergarten through second-grade class-rooms should focus on the following:

- A structure is in place for "playing" and working with materials and other students.

- Teachers clearly define the safety rules and procedures for using science equipment and materials.

- Students have access to a variety of science materials (e.g., centers, tubs of items, nonfiction and fiction science literature).

- Large- and small-group instruction is integrated throughout the day.

- Student work, with the science content, is displayed and changed with each unit focus.

- Students have an opportunity to read, write, and "do" science.

- The teacher reads aloud from science trade books or relevant literature.

- Students have an opportunity to share their experiences with science and the world around them.

- Time is provided for students to practice the science process skills (e.g., observing, measuring, inferring, communicating, predicting, and classifying).

- Students are allowed to experiment and report their findings.

- The teacher uses appropriate science vocabulary during instruction and checks for student understanding.

- There is a direct connection of life, earth, and physical science to the students' world and life outside of the classroom.

STEP 2: PROBLEM SOLVING

As teachers introduce the problem-solving process to students, keep in mind that consistency throughout the year will make it easier for students to complete the problems. A suggested instructional sequence is given below.

1. The teacher introduces the selected problem to the students.

2. Students, working alone or with a partner, work on a solution to the problem using writing or graphic interpretations to represent the solution.

3. Students are asked to share their ideas about how they solved the problem. This should include the related content and science process skills.

4. The teacher records some of the ideas on chart paper (this is the class data sheet).

5. The teacher has a quick discussion with the class to determine the best solution.

6. Students are asked to write several sentences, on their own, which describe the solution to the problem.

As the year progresses, the teacher should move from teacher-directed modeling to a whole class to doing the problem-solving task together. At this point, the students will record their thinking before the teacher has a class discussion. The goal is to have students make their own data sheet and write the solution on their own.

Problem-Solving Task

This particular problem is aligned to the second-grade standard of Sound and is aligned to the conceptual unit described in Step 3.

You are going to design a new instrument for the school band. The teacher gives you a set of materials and asks you to create an instrument that will make sound. You must also be able to show how sound is made and that the frequency of sound can be changed.

STEP 3: CONCEPTUAL UNDERSTANDING UNIT FOR SECOND GRADE

Exhibit 8.1 is a sample Conceptual Understanding Unit for second grade matched to the standard of Sound. It is aligned with the problem-solving task, the weekly science review, and the sample science literacy strategies.

STEP 4: MASTERY OF SCIENCE INFORMATION

Science Reviews

The science review occurs daily at the beginning of the science block of instruction. Students should complete the problems independently using paper and pencil while the teacher makes modifications to ensure that all students are successful working on their own. For early grades, teachers may want to give one or two of the "blocks" per day rather than all six.

The problems can either be written on the board, prepared in advance as an individual worksheet for each student, or given verbally as a start to the instructional day.

There are three types of weekly science review templates. An example of a taxonomy-based template (Exhibit 8.2) is provided as related to this second-grade conceptual unit. The questions are designed to increase in complexity as students work through the six boxes. However, teachers have the discretion to use fewer boxes.

EXHIBIT 8.1: Example of a Conceptual Understanding Unit for Second Grade

Grade Level: Second Grade   **Conceptual Unit Focus: Sound**

Standards and Indicators Matched to Unit Focus:

Teachers will list and "unwrap" the full text of their standards and indicators from individual state or district documents. *Source*: North Carolina Public Schools (www.ncpublicschools.org), North Carolina Standard Course of Study for Science, Competency Goal 4 (4.01, 4.02, 4.03, 4.04, 4.05)

"Unwrapped" Concepts: Need to Know About Sound:

❏ Investigations (about sound)

❏ Appropriate technology

❏ Production of sound (by vibrating objects and air)

❏ Frequency of sound (altering the rate of vibration, size, and shape of instruments)

❏ Human ear can detect sound

❏ Instruments make sound (including human vocal cords)

"Unwrapped" Skills: Be Able to Do:

❏ Conduct   ❏ Show

❏ Use   ❏ Observe

❏ Demonstrate   ❏ Describe

Topic or Context (specific lessons, textbook pages, learning activities used during the unit):

❏ Science textbook

❏ Other resources as determined by the teacher

Science Process Skills:

❏ Observing   ❏ Inferring

❏ Communicating   ❏ Predicting

❏ Classifying

Priority Vocabulary:

❏ Sound   ❏ Volume

❏ Pitch   ❏ Loud

❏ Vibration   ❏ Quiet

EXHIBIT 8.1: Example of a Conceptual Understanding Unit for Second Grade

Grade Level: Second Grade   **Conceptual Unit Focus: Sound**

Big ideas:

1. Sound can travel through different materials.

2. Human ears are specialized structures designed to receive sound.

3. Sounds are produced by vibrating objects.

Essential Questions:

1. What is sound?

2. How are sounds made?

3. How can sounds be changed?

Materials Needed:

To be determined by the teacher

Writing Prompt(s):

• How does sound change?

• Imagine that you are in a place where there is no sound. Describe what it would be like and how you would have to deal with the situation.

• What would happen if you couldn't hear for a day?

• You just met an alien and want to tell it about the sounds of earth. What would you tell the alien?

• Find a diagram of the human ear and tell how sound is detected by humans.

End-of-Unit Assessment

Each state has instituted ways to hold school districts and schools accountable for teaching the standards to the students. Both formative and summative assessments are used to determine proficiency in core content subject areas. The end-of-unit assessment may include a variety of test items, including selected response, constructed response, and performance tasks. End-of-unit assessments in science must include the problem-solving step in conjunction with the science process skills. For this reason, *Five Easy Steps* recommends that grade-level teams of teachers design and implement end-of-unit assessments based on their curriculum standards, the needs of their students, and state and local mandates.

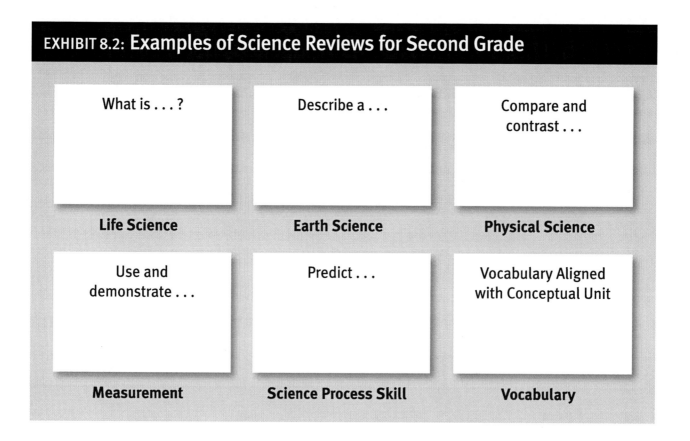

EXHIBIT 8.2: Examples of Science Reviews for Second Grade

What is . . . ?	Describe a . . .	Compare and contrast . . .
Life Science	**Earth Science**	**Physical Science**
Use and demonstrate . . .	Predict . . .	Vocabulary Aligned with Conceptual Unit
Measurement	**Science Process Skill**	**Vocabulary**

For the examples in Exhibit 8.2, you will have to insert the specific science strands for your individual district or school. The boxes contain topics that fall under each of the categories represented by the strands. It is suggested that you find or develop questions or problems related to your state objectives and benchmarks represented by each strand.

Using Graphic Organizers

This unit has an unlimited possibility of ideas for using graphic organizers. Exhibit 8.3 provides samples of graphic organizers that can be incorporated into the lessons and activities for Sound. Teachers should consider age-appropriate use of any of these in their classrooms.

Mastering Second-Grade Science Vocabulary

The early grades require that students learn science vocabulary related to the standards. Grade-level teams of teachers should identify the priority vocabulary related to the standards in the conceptual unit. Students should be presented with new vocabulary through visual, written, oral, and kinesthetic

EXHIBIT 8.3: Examples of Graphic Organizers

■ Compare and Contrast

Students could compare and contrast the pitch and frequency of different sounds or different instruments.

■ Sequence

Students could illustrate the process steps for showing how humans detect sound.

■ Main Idea and Details

After reading a story about sound, have students write or illustrate the main idea of the story.
Make sure that they provide supporting details.

■ Draw Conclusions

A fiction story about a world without sound could help students draw conclusions about the real or nonrealistic world.

■ Text Structure

An expository story selection about sound would show students the major text structures found in nonfiction literature. Students would identify the main text structures that are necessary for reading this type of text.

■ Cause and Effect

This graphic organizer would be used during a problem-solving activity. For example, "If the vibrations that produce sound change, what happens to the noise we hear?"

■ Problem–Solution

Students would be presented with a real-world problem and asked to provide a solution. Supporting details should be included with the solution.

activities. The science journal is a place where students can record their own interpretations of the vocabulary during the unit.

Chapter 4 provides several strategies for supporting vocabulary acquisition in second grade. Many of these activities could be recorded in the students' science journals.

STEP 5: COMMON FORMATIVE ASSESSMENTS

Common formative assessments for the conceptual unit should be designed by grade-level teams of teachers. It is important to remember that the common formative assessment be aligned only with the Priority Standards taught in the conceptual unit. Because there are so many variations in state standards and primary grade-level assessments, there is not a common formative assessment sample for this unit.

PART THREE

Resources for Implementation

Putting It All Together

CHAPTER 9

WITH A LITTLE BIT OF TIME AND PERSEVERANCE, you will be able to implement the *Five Easy Steps to a Balanced Science Program*.

TIME MANAGEMENT

It is a given fact that you will probably not be given any more time to teach science. In fact, with the intense concentration of reading and math across the country, many schools have reduced or abandoned the teaching of science altogether. Some science blocks in elementary school are done only twice a week within a thirty- to forty-five-minute time frame. This does not allow the students to move into any in-depth understanding of science, much less provide time for a hands-on lab or experiment. The time allocated for science instruction varies according to individual district and school mandates. Elementary school teachers have the ability to plan out their day but must work around set blocks of time for reading and math.

It is hoped that teachers will see the value of science and the real-world application of the process skills and problem solving. If there is a commitment to this program, then there is a way to incorporate all of the steps within a given time frame.

Scheduling the *Five Easy Steps*

Elementary science is often blocked into a specific time period during the day and typically is taught every other day or once or twice a week. The teacher must make a judgment call regarding how to best implement the *Five Easy Steps*. Many elementary teachers use the weekly science review two or three times a week rather than every day. This would be due to time constraints imposed by the literacy and math blocks. Assuming that the time period is fifty minutes, the teacher begins each lesson with the weekly science review. It is suggested that primary students focus on two to three of the areas in the beginning, rather than all six. Writing should be incorporated daily into all classrooms, and writing about science is appropriate during model writing or dictation. The conceptual unit, which would guide science instruction, identifies the priority vocabulary that should be introduced before the unit or topic is taught.

The literacy block lends itself to read-alouds and guided reading. Teachers who learn to balance nonfiction and fiction literature in science excite and engage their students in reading about science topics. The book *Brown Bear, Brown Bear: What Do You See?* is an excellent example of a nonfiction book that could lead to a discussion about real bears in the wild. *Cloudy with a Chance of Meatballs* is fun to read prior to introducing a unit on weather.

Grade-level teams of teachers should plan for instruction in science. Finding time to incorporate the *Five Easy Steps* takes time and deliberate planning. If you can incorporate one step at a time while working on the conceptual units, you may not be as challenged or frustrated with your teaching practices. It is also helpful to have a team of grade-level colleagues that can support you and brainstorm ideas with you.

Whatever your schedule, I hope that you will try something from the book. Kids love science, and once you try the steps, you will find that you can continually refine and reflect on these practices for months and years to come.

Best of luck for a great start to balancing your science program.

FREQUENTLY ASKED QUESTIONS

In all new programs, questions arise from the actual implementation of the steps. These questions are applicable to all grade levels and are included in all three editions of *Five Easy Steps to a Balanced Science Program*. Although the list in not inclusive, it gives a preliminary understanding of the most frequent questions about the steps.

How do I make my classroom look really inviting and "science-like"?

Every teacher has his or her own style of designing and creating inviting classrooms. The classroom should be clean, have relevant displays and student work, and provide areas where students can work individually or with small groups. Children are very visual and love the incorporation of color into the room. Many elementary teachers have a Science Center with materials and equipment so that students can explore and investigate science materials.

What are some activities to do during the first week of school to get my students comfortable with science?

The most important thing to do during the first weeks of school is to establish consistent rules and procedures for classroom management and organization. This is especially important in kindergarten, because some children have not experienced being in a school setting. The primary classroom is the perfect place for storytelling, sharing, and exploring the content. Some of the strategies in the book, including Science Scavenger Hunt, Who's in the Science Lab?, and The ABCs of Science, are great ways to get students involved with learning science.

What is more important: relationship building or rules and procedures?

There is nothing more important that establishing rules and procedures from day one. Teachers often combine the teaching of procedures with activities that allow students to get to know the teacher and each other.

What are good sources of science materials and equipment?

There are multiple science-supply companies that provide schools with commercial materials and science equipment. However, you do not need a large budget to effectively teach science in your classroom. Teams of grade-level teachers should go through their standards and decide on the hands-on activities that will be implemented within the conceptual unit. From this, the teachers create a list of supplies and materials, many of which are already in the school, that are needed to teach the selected labs and demonstrations. Local grocery stores, dollar stores, and home-improvement stores also have material that can be used for implementing effective science labs.

What do I do if I don't have a textbook?

I was quoted once to "just throw out the book," and although this was not a good thing to say at the time, I found the textbook to be a resource and not the curriculum. Teachers should use their state or district standards to design conceptual units, including the skills, concepts, vocabulary, and strategies that will allow students to go deeper into the science content.

What are the best ways to set up a lab station?

This depends on the arrangement of your classroom. Elementary teachers often have a Science Center that has manipulatives for students to use and explore. When teachers want to do a student lab, there are many ways to handle the materials, such as placing everything in a basket or on a tray. Some teachers have a Science Captain who is responsible for going to get the materials and bringing them back to their group. The best labs are those that are well planned, with the needs of the student addressed for the activity.

Where can I find good science problem-solving activities?

Once the conceptual units are designed by grade-level teams and teachers, then problem-solving activities can be created and implemented. There are many text resources and online sites that have a great selection of problem-solving experiments and activities. Some of these are found in the Webography section of this book.

How can I build my science competency, skills, and knowledge?

One of the best ways to become science competent is to read good science literature through current journals, online sites, and science organization Web sites. Being able to attend a local, state, or national science conference provides teachers with the most up-to-date information on materials, curricula, trends, and professional development. The National Science Teachers Association is the premier organization for supporting all grade levels of science educators and presents state, regional, and national conferences each year.

I just don't think I have time for problem solving, discrepant events, or creative challenge activities.

The AAAS and the National Science Education Standards clearly state that problem solving and the incorporation of the science process skills improve science literacy in students. If teachers will design conceptual units based on the Priority Standards and the "unwrapped" skills and concepts, then Step 2 of the *Five Easy Steps* becomes a key component of effective science teaching and learning.

There is too much content to cover. Any suggestions?

The format for designing a conceptual unit clearly delineates how to "cut down on the content" in science curricula. There is no additional time to teach, and with districts mandating more and more from teachers, we need to think on a critical level of what students really need to know and be able to do.

How many conceptual units should I design?

This depends on whether your district has mandates for the curriculum and pacing guides for the science content. It is suggested that you look at your required curriculum and "map out" the content that is the most important for your students to know.

Should I ever give science homework? What should it look like?

Yes, I give science homework. The grade level determines the amount of time that students would be asked to spend on homework. However, the homework must be relevant to the current unit of study and the student's life. The homework should tie into the skills taught in school and reflect lessons learned in class. These activities should focus specifically on the science process skills. Early-elementary students should have opportunities to observe, predict, communicate, infer, and measure the things around them. Many primary teachers create homework packets that include an integration of reading, writing, and the content areas. This could include reading aloud a science-related book, going on an outside adventure, or making predictions about the world around them.

How do you deal with the massive amount of vocabulary in science?

Vocabulary needs to be selected on the basis of the Priority Standards and be incorporated into the conceptual unit. Best instructional practices for literacy support the student's acquisition of this vocabulary. It is important for students to "See it," "Say it," and "Write it."

What is the best method for developing good inquiry questions?

Good questioning begins with the teacher understanding the levels of cognitive thinking and using these on a daily basis in the science lesson. Many teachers do not challenge their students with higher-level questions, causing them to be at a disadvantage during high-stakes testing.

How often do you have students use the science process skills?

The science process skills are the backbone of good scientific thinking and processing. Teachers should make sure that the process skills are being used every day.

As a busy, self-contained elementary teacher having to teach all students, how can I incorporate literacy and math within the science content to help me "kill two birds with one stone?"

Science is a subject that can be integrated with every other content. Students love to hear stories about the world around them, and incorporating good science literature into the conceptual unit encourages interest, curiosity, and motivation for learning science information. I always bring in both fiction and nonfiction literature selections to introduce my students to the science concepts. Topical units, such as animals, the ocean, or the senses, have a multitude of good texts for exposing students to science. Thematic units on change, conflict, and relationships can also be incorporated using children's literature. One of my favorite student teachers used the book *Jurassic Park,* along with nonfiction information on dinosaurs, to introduce the unit on geologic time. She read the book aloud to the students, created engaging lab activities, had students assume the role of a paleontologist, and had students record their adventures in their science journals. Chapters 3 and 4 have several suggestions for reading, writing, and vocabulary skill development in science. As always, teachers must choose age-appropriate reading materials for the students.

The standards for my state are very content specific. I want to teach conceptually but have limited time. What suggestions do you have to make sure that students are prepared for the state assessments, which are concept and vocabulary heavy?

It seems that most state science standards are very content specific and heavy with details. I suggest that teachers take the state science curriculum documents and identify the Priority Standards that address the concepts that students must know and the skills that students must be able to do. This is not to say that the other standards are of less value, but this allows teachers to teach the ones that carry the most weight for the state assessments. It is very important that every teacher incorporate the problem-solving steps at least every two weeks and the science process skills every day. This can be done through warm-up or bell activities, guided and independent practice activities, and summarizing at the end of the day. Teachers should also identify the priority vocabulary that matches the conceptual unit and use the strategies that support reading, writing, and vocabulary acquisition.

Are there specific textbooks that you recommend over others for supporting *Five Easy Steps* Science?

I have always thought that the science textbook should be used as a reference and not the curriculum. When teachers design and implement conceptual units in science, the Priority Standards should drive the instruction. The "unwrapped" concepts and skills determine what the student should know and be able to do and serve as the basis for the instructional planning and assessment. The textbook can also be used to teach students about informational text structures and teach them to identify main ideas and the author's purpose. There are several strategies in Chapter 3 that address pre-, during-, after-reading text selections. These are excellent to use with students as they read the textbook. Science has a tremendous amount of vocabulary that is often new and confusing to students. Using effective vocabulary strategies, with the textbook, allows students to learn and comprehend new terms in science content.

One suggestion is to pair a fiction book with a nonfiction book in science. Students will be exposed to good literature in addition to informational text at the same time. I always had a "science reading center" in my classroom with a variety of literature selections. Students could spend time exploring or investigating different topics of interest during independent reading time or after completing assignments. I also had students select one topic of interest during a quarter grading period. They would read, research, and write their own book on the topic and present it to the class. One of my favorites was a book entitled *The Bats and Me*. It included facts about bats, an original poem, drawings and illustrations, and a self-reflection of what was learned.

From an elementary-grade-level perspective, I struggle with getting my teachers to be able to work together to build assessments that address skills that are common to all disciplines.

Teams of teachers that collaboratively design, administer, score, and analyze are provided with valid feedback regarding their students' current understanding

EXHIBIT 9.1: A Traditional Science Classroom Compared to Rethinking Science Instruction

Traditional Science Classroom	Rethinking Science Instruction
• One instructional arrangement	• Multiple arrangements of students
• Teacher-directed instruction	• Student-centered learning
• Few manipulatives or hands-on materials	• Multiple manipulatives and hands-on materials
• One lab per unit	• Labs or minilabs related to the conceptual unit
• Whole-class instruction predominates	• Mixture of whole class, cooperative groups, and individual opportunities for learning
• Inflexible block of time	• Flexible time groupings
• Single teacher planning units/topics	• Grade-level/content teams plan conceptual units of instruction
• Assessment OF learning	• Assessment FOR learning
• Single definition of "proficient"	• Multiple opportunities to be "proficient" or "exemplary"
• Single text or textbook	• Multiple resources/text/Internet access
• Teacher does the problem solving	• Problem-solving task and write-up completed by the student
• Whole-class standards for grading	
• Single form of assessment	• Multiple forms of assessment (i.e., selected, constructed, performance)
• Data analysis is not used to adjust instruction	• Data analysis by Data Teams guides instruction and learning
• No or little emphasis on science vocabulary	• Daily emphasis on science vocabulary
• No or little integration of reading strategies	• Daily emphasis on strategic reading
• No or little integration of writing/journaling	• Daily emphasis on writing/journaling
• No or little integration of visual and nonlinguistic strategies	• Nonlinguistic representations used by students

and knowledge of the Priority Standards. There must be a concerted effort to arrange time for teachers to plan together during the week in order for this work to happen.

What are the most important questions to ask myself in order to implement the *Five Easy Steps to a Balanced Science Program*?

I suggest using Exhibit 9.2 to address the steps in implementing the *Five Easy Steps*. It is important to look for and reflect on the evidence that supports the application of all components within the steps. For example, Step 4, Mastery of Science Information, contains a wide range of writing strategies and activities for the students. You would want to hold yourself accountable for using the best practices during this step.

A FRAMEWORK FOR IMPLEMENTATION

Rethinking Science Instruction

School leaders and planning teams may use the following information and Exhibit 9.2 for reviewing the current state of their science program. In a time when science is moving forward into state testing, it is imperative for schools to be proactive in planning and implementing a balanced science program. Exhibit 9.1 compares the traditional science classroom to a plan whereby teachers are rethinking science instruction.

Getting Started with the *Five Easy Steps to a Balanced Science Program*

Before you begin to review and critique your current science program, take time to examine the chart in Exhibit 9.2. As a team, reflect on what the current characteristics are within your district or school. Once you have determined your school's needs, use the templates in Exhibit 9.3 to help plan your balanced science program.

Planning Templates for the *Five Easy Steps*

The templates presented in Exhibit 9.3 should be used as a strategic planning tool for analyzing the *Five Easy Steps to a Balanced Science Program*. In Exhibit 9.3, the first column represents the desired state if the *Five Easy Steps* program is implemented with fidelity and completeness. The middle column is a reflection of the present science program. To analyze this effectively, a team

EXHIBIT 9.2: Current Characteristics of the Science Program in the School or District

STEP 1: Establishing an Effective Science Environment   EVIDENCE

	EVIDENCE
What characteristics of my classroom represent an effective science environment?	
What strategies am I using to promote a positive learning environment for science education?	
How have I implemented safety rules and procedures in my science classroom?	
How do I incorporate lab experiences for my students?	
What classroom management strategies are most effective for me and my teaching style?	
How do I differentiate instruction to meet the needs of my students?	

EXHIBIT 9.2: Current Characteristics of the Science Program in the School or District

STEP 2: Problem Solving	EVIDENCE
How are the problem-solving tasks aligned with the Priority Standards?	
How do I incorporate problem solving in my classroom every two weeks?	
How do I incorporate the problem-solving data sheet and task write-up templates?	
What discrepant events have I used in my teaching?	
How are the problem-solving scoring guides designed and implemented?	
How are students involved with their self-assessment in the problem-solving tasks?	

EXHIBIT 9.2: Current Characteristics of the Science Program in the School or District

STEP 3: Conceptual Understanding	EVIDENCE
What conceptual units have I designed that are aligned with my science standards?	
How have I collaborated with my team members on conceptual units?	
How do I use questioning in my classroom?	
What strategies have I implemented to ensure effective questioning?	
How am I implementing reading aloud in science?	
What pre-reading strategies have I used with science content?	
What during-reading strategies have I used with science content?	
What post-reading strategies have I used with science content?	

EXHIBIT 9.2: Current Characteristics of the Science Program in the School or District

STEP 4: Mastery of Science Information	EVIDENCE
How am I implementing a weekly science review?	
How do I incorporate the four types of writing in my classroom each week?	
What kinds of writing prompts have my students used?	
How are my students using their science journals?	
What vocabulary strategies do I incorporate into my lessons each week?	
How do I have my students use graphic organizers each week?	

EXHIBIT 9.2: Current Characteristics of the Science Program in the School or District

STEP 5: Common Formative Assessments	EVIDENCE
How has my team used the common formative assessment process?	
What types of common formative assessment items have my students seen and completed each week?	
How has my team analyzed the results of our common formative assessments?	

ADDITIONAL THOUGHTS	EVIDENCE
How am I continuing my learning and professional development in science?	
What resources do I use most for teaching science?	
What other resources do I need to be more effective?	

of educators should honestly and openly reflect and respond to each individual component represented in the first column. The last column provides space for implementation steps.

Prior to filling in the template in Exhibit 9.3, teams should reflect on and respond to the successes, weaknesses, and challenges of science instruction within their school, grade level, or individual classroom. These guiding thoughts will enable the team to better understand the necessary implementation steps.

In addition, templates for month-by-month planning are included in Exhibits 9.4 through 9.6. It is recommended that the implementation of the steps take time but with a focus of 100 percent within three years.

I sincerely hope that these templates will help schools and districts made appropriate decisions about improving the science program.

Throughout the implementation process, it is important to look for and reflect on the evidence that supports the application of each component within each step. For example, Step 4, Mastery of Science Information, offers a wide range of writing strategies and student activities. You would want to monitor how each strategy affects student achievement and hold yourself accountable for using the best strategies.

EXHIBIT 9.3: Strategic Planning for Analyzing the *Five Easy Steps*

STEP 1: Establishing an Effective Science Environment

Strengths: _____

Weaknesses: _____

Challenges: _____

Desired Results for *Five Easy Steps*	Current Analysis of Science Program	Implementation Steps
I feel comfortable teaching science.		
Students understand the definition of "science" and the role of a scientist.		
Teachers reflect on the characteristics of effective science teaching.		
Relationship strategies are identified and used by teachers.		
Teachers understand and model the appropriate use of science tools, equipment, and substances.		
All classrooms post and consistently enforce all science safety rules and procedures.		
All students and parents or guardians have signed and returned the science safety contract.		
Science labs are aligned with the course standards.		
Appropriate lab formats are used at each grade level.		

EXHIBIT 9.3: Strategic Planning for Analyzing the *Five Easy Steps*

STEP 2: Problem Solving

Strengths: _____

Weaknesses: _____

Challenges: _____

Desired Results for *Five Easy Steps*	Current Analysis of Science Program	Implementation Steps
Problem-solving strategies are identified and used by all teachers.		
Problem-solving strategies are taught to students.		
Problem-solving tasks are aligned with the course standards in the conceptual unit.		
Grade-level and content teams collaborate on the problem-solving task.		
Problem solving occurs at least twice a month.		
Problem-solving scoring guides are designed and used by grade-level and content teams.		
Students are involved in a self-assessment of their problem-solving task.		
Resources are provided to teachers with ongoing support as needed.		
Discrepant events are demonstrated on a monthly basis.		

EXHIBIT 9.3: Strategic Planning for Analyzing the *Five Easy Steps*

STEP 3: Conceptual Understanding

Strengths: _____

Weaknesses: _____

Challenges: _____

Desired Results for *Five Easy Steps*	Current Analysis of Science Program	Implementation Steps
Conceptual units are designed by grade-level and content teams.		
Conceptual units are aligned with the state and district curricula and testing information.		
Priority Standards and "un-wrapped" skills and concepts are emphasized in the conceptual unit.		
Big Ideas and Essential Questions are used to guide instruction and assessment.		
Science vocabulary and process skills are identified and incorporated into the conceptual unit.		
Effective questioning strategies are included and used with teacher-directed instruction.		
Science literacy is emphasized at all grade levels.		
Best instructional practices for reading in science are used.		

EXHIBIT 9.3: Strategic Planning for Analyzing the *Five Easy Steps*

STEP 4: Mastery of Science Information

Strengths: _____

Weaknesses: _____

Challenges: _____

Desired Results for *Five Easy Steps*	Current Analysis of Science Program	Implementation Steps
A weekly science review is used in all grades daily.		
Students use graphic organizers to demonstrate understanding of science concepts.		
My team has identified priority vocabulary for each conceptual unit.		
My team collaborates on planning the weekly science review problems or questions.		
Science vocabulary strategies are integrated into daily instruction.		
Writing prompts are used in every grade weekly.		
Science notebooks and/or journals are used by students on a daily basis.		
Resources for weekly science reviews, vocabulary strategies, and writing are available to teachers.		

EXHIBIT 9.3: Strategic Planning for Analyzing the *Five Easy Steps*

STEP 5: Common Formative Assessments

Strengths: _____

Weaknesses: _____

Challenges: _____

Desired Results for *Five Easy Steps*	Current Analysis of Science Program	Implementation Steps
Grade-level and content teams collaboratively design science assessments FOR learning.		
Common formative assessments are aligned to the conceptual unit (Priority Standards and "un-wrapped" skills and concepts).		
Common formative assessments include selected-response and constructed-response items.		
Grade-level and content teams score the common formative assessments for science.		
Student data are analyzed and acted upon by grade-level and content teams.		
Assessment results are used to guide instruction.		
Assessment results are posted on a data wall.		
Differentiation of instruction occurs as a result of data analysis.		
Constructed-response items and performance tasks have a scoring guide designed by collaborative teams.		

EXHIBIT 9.4: Time Frame for Implementation of *Five Easy Steps*: Year 1

Implementation Evidence

July	
August	
September	
October	
November	
December	
January	
February	
March	
April	
May	
June	

EXHIBIT 9.5: Time Frame for Implementation of *Five Easy Steps*: Year 2

Implementation Evidence

July	
August	
September	
October	
November	
December	
January	
February	
March	
April	
May	
June	

EXHIBIT 9.6: Time Frame for Implementation of *Five Easy Steps*: Year 3

Implementation Evidence

July	
August	
September	
October	
November	
December	
January	
February	
March	
April	
May	
June	

Reproducibles

APPENDIX A

T HIS APPENDIX INCLUDES TEMPLATES AND CHARTS that will be of assistance to teachers in implementing the *Five Easy Steps to a Balanced Science Program.* Teachers have permission to copy these for instructional purposes only.

The following reproducibles are included:

- **Step 1: Establishing an Effective Science Classroom.** Five templates (Exhibits A.1 through A.5) are provided that match the instructional strategies found in Step 1. Teachers may want to reproduce these for the students or use them on the overhead or smartboard for class discussion.

- **Step 2: Problem Solving.** A template (Exhibit A.6) is provided for the Problem-Solving Task Data Sheet for Students (Primary Grades). There are also sample creative challenge activities (Exhibit A.7) in this step.

- **Step 3: Conceptual Understanding.** The template for the conceptual unit used in Step 3 (Exhibit A.8) is found here. Templates for the questioning cube, the wheel, and the graphic organizer (Exhibits A.9, A.10, and A.11) are provided. The cooperative learning roles for science are included in student-friendly templates (Exhibits A.12 through A.17) in this section.

- **Step 4: Mastery of Science Information.** Three sample weekly science reviews are provided for this step: (1) standards-based, (2) content-based, and (3) taxonomy-based (Exhibits A.18 through A.20).

STEP 1: ESTABLISHING AN EFFECTIVE SCIENCE CLASSROOM

EXHIBIT A.1: Who's in the Science Lab?

EXHIBIT A.2: Science Scavenger Hunt

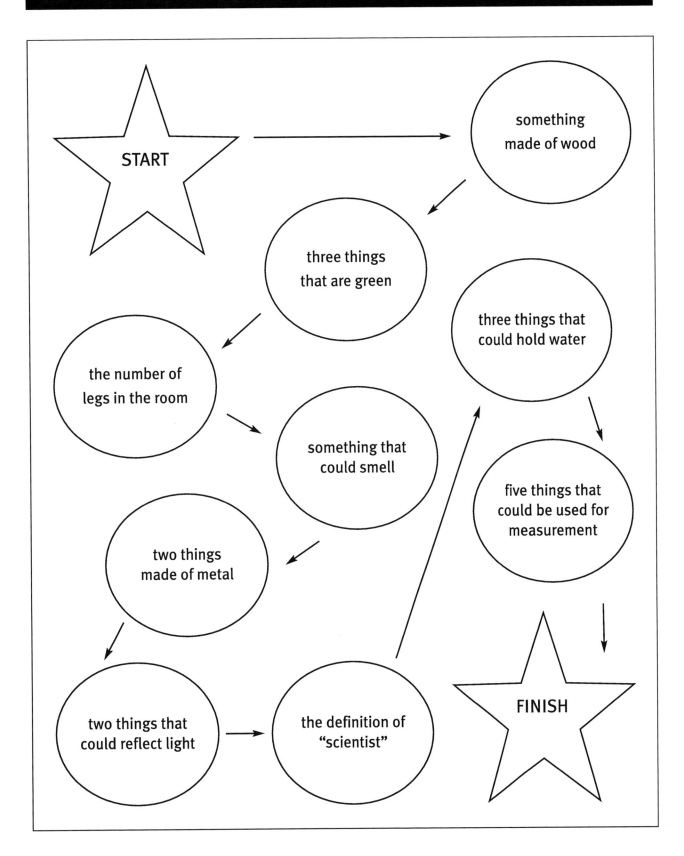

EXHIBIT A.3: The ABC's of Science

A	B	C	D	E
F	G	H	I	J
K	L	M	N	O
P	Q	R	S	T
U	V	W	X	Y
Z				

EXHIBIT A.4: Equipment Bingo

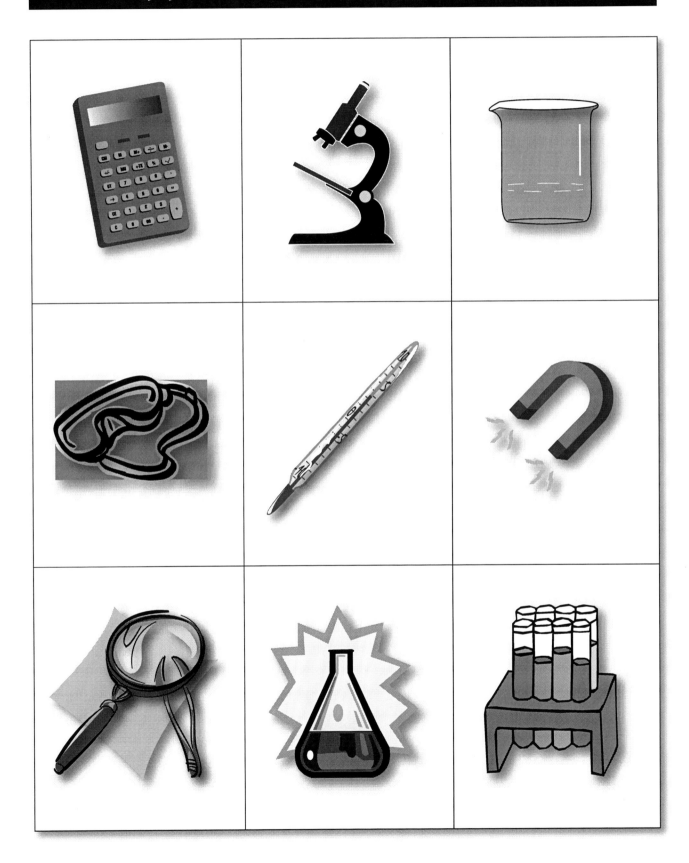

EXHIBIT A.5: Video Tic-Tac-Toe

Fact	One Thing I Already Know	Fact
One Question	TOPIC	One Interesting Observation
Fact	One Thing My Friend Knows	Fact

STEP 2: PROBLEM SOLVING

EXHIBIT A.6: **Problem-Solving Task Data Sheet for Students: Primary Template**

Name: **Date:**   **Title of Problem:**

Important Science Vocabulary:

Hypothesis (What I Think):

Steps to the Solution:

What Actually Happened:

My Understanding:

Solving the Problem:

1. Use your favorite strategy to solve the problem.

2. Number or label your steps.

3. Write a sentence or two telling how you solved the problem.

4. Go back and check to make sure that you are finished.

Problem Solving Write-Up:

Use the problem-solving write-up template to record your information.

Make sure that you include science vocabulary in your report.

EXHIBIT A.7: Sample Creative Challenge Activities

Creative Challenge: Stepping Through Paper

The Challenge:

How can you make a hole in the piece of paper large enough to step through?

Materials:

One piece of notebook paper

Solution:

1. Fold the paper in half

2. Cut two slits near the right and left edges of the paper, each perpendicular to the fold. The cuts must come down from the folded half, not up from the unfolded edges.

3. Cut off the folded edge from slit to slit. Be sure not to snip off the ends of the paper.

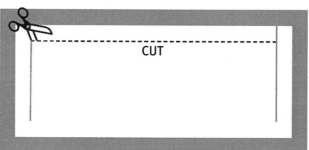

4. Make about forty cuts with your scissors. Alternate each cut, first coming down from the folded edge and then coming up from the unfolded edges. The cuts should be parallel to the first slits you made.

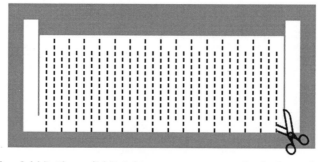

5. Open up the paper and unfold it. If you did it right, you can put your body through the opening.

EXHIBIT A.7: Sample Creative Challenge Activities

Creative Challenge: The Leaping Ping-Pong Ball

The Challenge:

How can you get the Ping-Pong ball from one cup to the other without touching the cups or the Ping-Pong ball?

Materials:

Two empty plastic cups

One Ping-Pong ball

Solution:

This problem works on the Bernoulli Principle. You must practice this first. Place the cups about three inches apart. Place the Ping-Pong ball in one of the cups. Blow at an angle to "lift" the Ping-Pong ball out of the first cup and into the second cup.

EXHIBIT A.7: Sample Creative Challenge Activities

Creative Challenge: Paper Tower

The Challenge:

How can you create the tallest free-standing tower using only one piece of paper and the materials provided?

Materials: Per group of students

One piece of 8 1/2 x 11 paper

Three-inch masking tape

Four paper clips

Teacher material:

One-meter stick

Solution:

There is not a clear solution because each group will have a variety of designs.

EXHIBIT A.7: Sample Creative Challenge Activities

Creative Challenge: Float a Boat

The Challenge:

How can you devise a contraption that will hold the most weight when it is floating in water?

Materials: Per group of students

One piece of card stock

Four paper clips

Two paper cups

Four craft sticks

Four drinking straws

One balloon

Teacher Materials:

One hundred or more pennies (the weights)

One water-filled container

Solution:

There is no one solution for this challenge. Carefully place the weights on the boat until it sinks or flips over.

EXHIBIT A.7: Sample Creative Challenge Activities

Creative Challenge: Four-Item Balance

The Challenge:

How can you devise a contraption that will balance the four items on the glass?

Materials: Per group of students

Two paper cups

One fork

One drinking glass

One metal spoon

One drinking straw

Solution:

There is no one solution for this challenge.

EXHIBIT A.7: Sample Creative Challenge Activities

Creative Challenge: Aloft in the Air

The Challenge:

How can you assemble a device that will remain aloft the longest when it is dropped?

Materials: Per group of students

One drinking straw

Two craft sticks

One rubber band

One 3" x 5" index card

One three-inch piece of tape

One balloon

Teacher materials:

Stopwatch

Solution:

There is no one solution for this challenge.

EXHIBIT A.7: Sample Creative Challenge Activities

Creative Challenge: The Catapult

The Challenge:

How can you create a device that will catapult a ball of cotton the longest distance?

Materials: Per group of students

Two craft sticks

Four paper clips

Four paper cups

Twelve inches of tape

Four rubber bands

One cotton ball

One piece of 8 1/2 x 11 paper

Twelve inches of string

One cardboard tube (toilet paper roll)

Solution:

There is no one solution to this challenge.

EXHIBIT A.7: Sample Creative Challenge Activities

Creative Challenge: Trash Can Olympics

The Challenge:

How can you create a device that will move five marbles to the trash can over a distance of five feet?

Materials: Per group of students

One-meter stick

Twelve inches of tape

Ten paper clips

Five marbles

One cardboard tube

Two balloons

One ten-inch piece of string

Three rubber bands

One object of student choice

Teacher Material:

One trash can

Solution:

There is no one solution to this challenge.

EXHIBIT A.7: Sample Creative Challenge Activities

Creative Challenge: Making Noise

The Challenge:

How can you create a device that will make the most noise (as determined by audience vote)?

Materials: Per group of students

Two craft sticks

Four paper clips

Paper cups

Twelve inches of tape

One twelve-inch piece of string

Three marbles

Two rubber bands

One cup of water

One balloon

Three straws

Solution:

There is no one solution to this challenge.

EXHIBIT A.7: Sample Creative Challenge Activities

Creative Challenge: Above the Floor

The Challenge:

How can you devise a contraption that will balance the five pennies the longest on the fishing line suspended above the floor?

Materials: Per group of students

Three craft sticks

Four paper clips

Four paper cups

Twelve inches of tape

One two-liter plastic bottle

Five pennies

One balloon

One object of student choice

Teacher Materials:

A piece of fishing line strung across the room above the floor

Solution:

There is no one solution to this challenge.

STEP 3: CONCEPTUAL UNDERSTANDING

EXHIBIT A.8: The Conceptual Unit Design Template

Grade Level: **Conceptual Unit Focus:**

Standards and Indicators:

"Unwrapped" Concepts (Need to Know About):

"Unwrapped" Skills (Be Able to Do):

Science Process Skills:

Priority Vocabulary:

EXHIBIT A.8: The Conceptual Unit Design Template

Grade Level: **Conceptual Unit Focus:**

Big Ideas:

1.

2.

3.

Essential Questions:

1.

2.

3.

Resources:

(specific lessons, textbook pages, and learning activities to be used during the unit)

Materials:

Time Frame:

EXHIBIT A.8: The Conceptual Unit Design Template

Grade Level: **Conceptual Unit Focus:**

Discrepant Event(s):

Writing Prompt(s):

Problem-Solving Task:

End-of-Unit Assessment:

Selected Response:

Constructed Response:

Performance Tasks:

Instructional Strategies:

EXHIBIT A.8: The Conceptual Unit Design Template

Scoring Guide:

Exemplary:

All "proficient" activities PLUS:

Proficient:

Progressing:

Beginning:

EXHIBIT A.8: The Conceptual Unit Design Template

Peer Evaluation:

Self-Evaluation:

Teacher Evaluation:

Comments:

EXHIBIT A.9: Questioning Cube

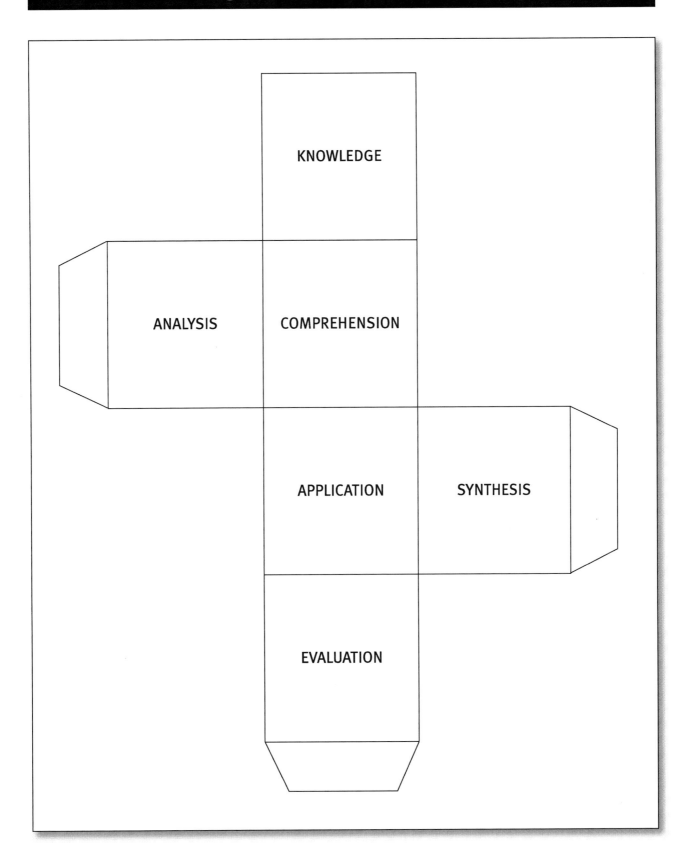

EXHIBIT A.10: Questioning Wheels

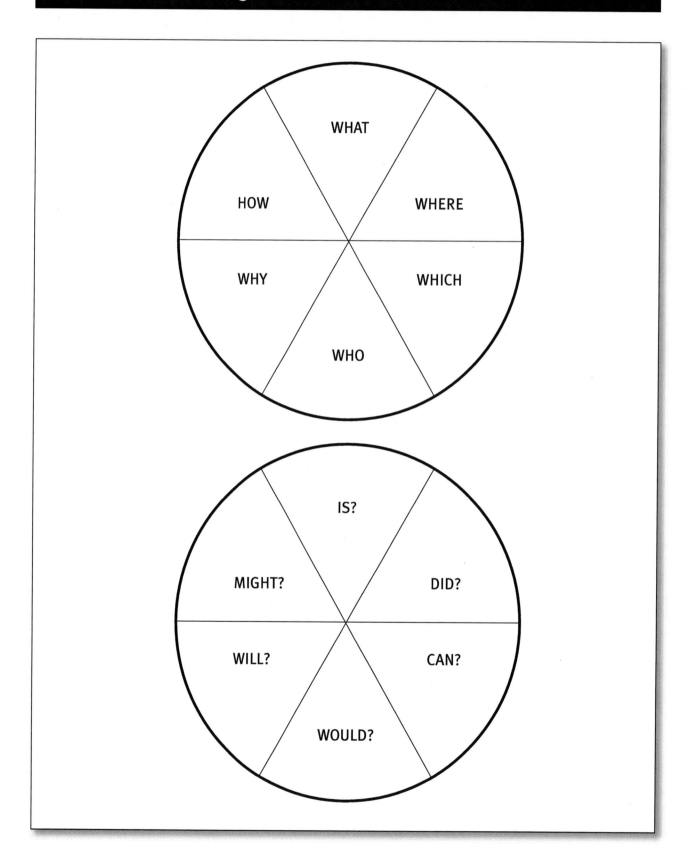

EXHIBIT A.11: Questioning Organizer

Bloom's Taxonomy

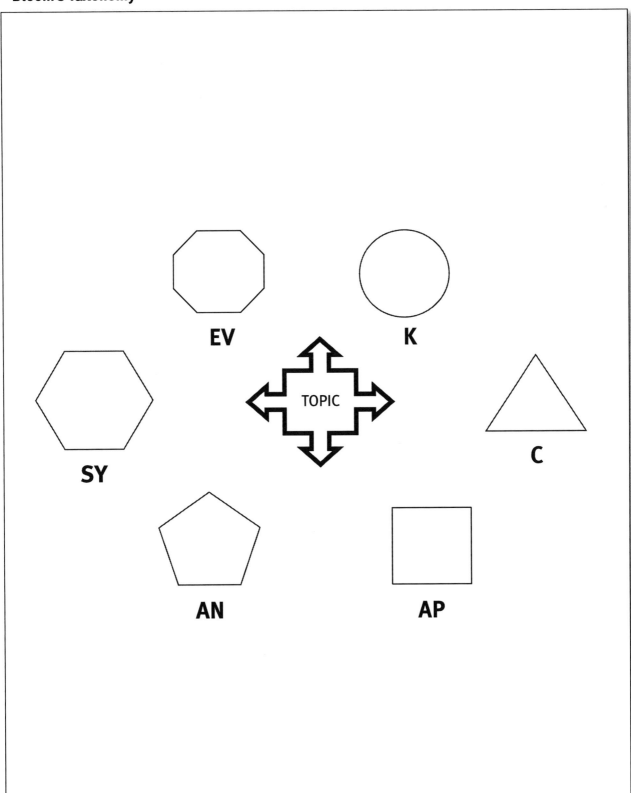

EXHIBIT A.12: BE A SCIENTIST: Creative Connector

Cooperative Learning Role: Creative Connector

Name:

Text Selection:

Date:

Assignment: (p. _____ to p. _____)

Role: Creative Connector

Your job is to find connections between the literature and the world. This includes connecting the reading to your own life, to events at school and in the community, to outside problems, and to daily events. Make connections between this literature and other writings on the same topic or by the same author.

Write your own personal connection and ask each member to make a connection. Use the response questions as a guide.

My Personal Connections:

Questions to Guide Group Members to Make Connections:

EXHIBIT A.13: BE A SCIENTIST: Discussion Director

Cooperative Learning Role: Discussion Director

Name:

Text Selection:

Date:

Assignment: (p. _____ to p. _____)

Role: Discussion Director

Your job is to develop a list of questions that your group might want to discuss about this content. Don't worry about details. Help your group discuss the main ideas and share their reactions. The best discussion comes from your own thoughts, but you may use some of the question cues to develop your discussion.

Possible questions or topics for today:

1.

2.

3.

Possible questions to ask:

What were your thoughts as you read the text selection(s)?

How did you feel while reading this selection?

What questions do you need to clarify?

What interested you in the reading?

EXHIBIT A.14: BE A SCIENTIST: Literary Luminary

Cooperative Learning Role: Literary Luminary

Name:

Text Selection:

Date:

Assignment: (p. _____ to p. _____)

Role: Literary Luminary

Your job is to develop a list of questions that your group might want to discuss about this content. Don't worry about details. Help your group discuss the main ideas and share their reactions. The best discussion comes from your own thoughts, but you may use some of the question cues to develop your discussion.

Location in Text:

page _____ paragraph _____

page _____ paragraph _____

page _____ paragraph _____

page _____ paragraph _____

Characteristics of the text:

Important

Informative

Surprising

Controversial

Humorous

Confusing

Well written

EXHIBIT A.15: BE A SCIENTIST: Illustrator

Cooperative Learning Role: Illustrator

Name:

Text Selection:

Date:

Assignment: (p. _____ to p. _____)

Role: Illustrator

Your job is to draw a visual related to the reading. It may be a sketch, cartoon, diagram, or graphic organizer. You may draw something that's discussed or something that made a connection in the reading. You are to share your illustration before asking for group feedback.

Presentation of the illustration:

As the discussion is underway, complete your illustration. Share your thoughts about your work before asking for feedback. One at a time, ask members to hypothesize what the illustration represents. Then share your ideas. Ask for feedback and adjust the illustration, if needed.

Possible questions to address in the illustration:

What if . . . ?	Which would . . . ?
What might have happened . . . ?	Who and where . . . ?
What sequence of . . . ?	Imagine that . . .
Why did . . . ?	The best answer . . .
How would . . . ?	The reason is . . .

EXHIBIT A.16: BE A SCIENTIST: Summarizer

Cooperative Learning Role: Summarizer

Name:

Text Selection:

Date:

Assignment: (p. _____ to p. _____)

Role: Summarizer

Your job is to prepare a brief summary of the literature read today. Make your comments one to two minutes in duration, focusing on the key concepts and ideas. List the key points below. You may use any strategy for helping your group to remember the important components.

Summary:

Key points for discussion:

EXHIBIT A.17: BE A SCIENTIST: Word Wizard

Cooperative Learning Role: Word Wizard

Name:

Text Selection:

Date:

Assignment: (p. _____ to p. _____)

Role: Word Wizard

Your job is to identify vocabulary words or phrases that the group finds challenging or hard to understand. After reading the text, select three to five words or phrases that need discussion. Based on the context clues, guess the meaning of the word, look it up, and be ready to share with the group.

Word Wizard words:

Strategies to help my group understand the vocabulary:

STEP 4: MASTERY OF SCIENCE INFORMATION

EXHIBIT A.18: Weekly Science Review: Standards-Based Template

Big Ideas	Content/Concept
Vocabulary	**Science in the World**
Science Process Skill(s)	**Measurement/Graphing**

EXHIBIT A.19: Weekly Science Review: Content-Based Template

Earth and Space Science	Physical Science
Environmental Science	**Life Science**
Science Process Skill	**Vocabulary**

EXHIBIT A.20: Weekly Science Review: Taxonomy-Based Template

Knowledge	Comprehension
Application	**Analysis**
Synthesis	**Evaluation**

References

Ainsworth, L., & J. Christinson. 2006. *Five easy steps to a balanced math program.* Englewood, CO: Lead + Learn Press.

Ainsworth, L., & D. Viegut. 2006. *Common formative assessments: How to connect standards-based instruction and assessment.* Thousand Oaks, CA: Corwin Press.

American Association for the Advancement of Science. 1989, 1990. *Science for all Americans: A project 2061 report on literacy goals in science, mathematics and technology.* Washington, D.C.: AAAS.

American Association for the Advancement of Science. 1993. *Benchmarks for science literacy.* New York: Oxford University Press.

Anderson, L. W., & D. R. Krathwohl (eds.). 2001. *A taxonomy for learning, teaching and assessing: A revision of Bloom's taxonomy of educational objectives.* Complete edition. New York : Longman.

Anderson, R. C., et al. 1985. *Becoming a nation of readers: The report of the commission on reading.* Washington, D.C.: The National Institute of Education.

Barkley, E. F., K. P. Cross, & C. H. Major. 2005. *Collaborative learning techniques: A handbook for college faculty.* San Francisco, CA: Jossey-Bass

Black, P., & D. Wiliam. 1998. Assessment and classroom learning. *Assessment in Education: Principles, Policy, and Practice,* 5(1): 7–73.

Bloom B. S. 1956. *Taxonomy of educational objectives, handbook I: The cognitive domain.* New York: David McKay.

Bransford, J. D., A. L. Brown, & R. R. Cocking. 1999. *How people learn: Brain, mind, experience and school.* Washington, D.C.: National Academies Press.

Center for Science, Mathematics, and Engineering Education. 1998. *Every child a scientist: Achieving science literacy for all.* Washington, D.C.: National Academies Press.

Cocker, E. 1730. *Arithmetick.* London: Edw. Hamilton and Sam Fuller.

Connecticut Department of Education. Connecticut Prekindergarten–Grade 8 Science Curriculum Standards Including Grade-Level Expectations 1.2 (1.2.1, 1.2.2, 1.2.3, 1.2.4). Available at www.ct.gov/sde.

Cunningham, P. M., & R. L. Allington. 2007. *Classrooms that work: They can all read and write.* 4th ed. Boston, MA: Allyn and Bacon.

Cutlip, G. W. (series ed.). 2005. *Balanced assessment: The key to accountability and improved student learning.* Washington, D.C.: National Education Association.

Data teams. 2008. The Leadership and Learning Center. Englewood, CO: Lead + Learn Press.

Duke, N. K. 2004. The case for informational text. *Educational Leadership* 61 (6): 40–44.

Einstein, A. 2010. ThinkExist.com. www.ThinkExist.com.

Flack, J. D. 1993. *TalentEd: Strategies for developing the talent in every learner.* Englewood, CO: Teacher Ideas Press.

Gabel, D. Winter. 2003. Conceptual understanding of Science. *Educational Horizons,* 81 (2): 70–76.

Hubble, E. P. 1954. *The nature of science, and other lectures.* San Marino, CA: Huntington Library.

Johnson, D. W., & R. T. Johnson. 1989. *Cooperation and competition: Theory and research.* Edina, MN: Interaction Book Company.

Marzano, R. 1988. *Dimensions of thinking: A framework for curriculum and instruction.* Alexandria, VA: ASCD.

Marzano, R. 2003. *What works in schools: Translating research into action.* Alexandria, VA: ASCD.

Marzano, R., D. Pickering, & J. Pollock. 2001. *Classroom instruction that works.* Alexandria: VA: ASCD.

Mathematics and Engineering Staff Center for Science. 1998. *Every child a scientist: Achieving science literacy for all.* Washington, D.C.: National Academies Press.

McTighe, J., & G. Wiggins. 2005. *Understanding by design.* Alexandra, VA: ASCD.

Mead, M., & R. Métraux. 1957. Image of the scientist among high school students: A pilot study." *Science,* 126:386–387.

Merriam-Webster Dictionary Online, 2010. Merriam-Webster Incorporated.

National Academy of Sciences. 1996. *National science education standards.* Washington, D.C.: National Academies Press, pp. 11, 13–15, 104–106.

National Academy of Sciences. 2000a. *Inquiry and the national science education standards: A guide for teaching and learning.* Committee on the Development of an Addendum to the National Science Education Standards on Scientific Inquiry. S. Olson and S. Loucks-Horsley, eds. Washington, D.C.: National Academies Press.

National Academy of Sciences. 2000b. *National science education standards: A guide for teaching and learning.* Washington, D.C.: National Academies Press.

Nelson, G. 1998. Science literacy for all: An achievable goal?" *Optical Society of America,* 9 (9): 42.

North Carolina Public Schools. North Carolina Standard Course of Study for Science, Competency Goal 2 (2.01, 2.02). Available at www.ncpublicschools.org.

North Carolina Public Schools. North Carolina Standard Course of Study for
    Science, Competency Goal 2 (2.01, 2.02, 2.03, 2.04, 2.05). Available at
    www.ncpublicschools.org.

North Carolina Public Schools. North Carolina Standard Course of Study for
    Science, Competency Goal 4 (4.01, 4.02, 4.03, 4.04, 4.05). Available at
    www.ncpublicschools.org.

Poincare, H. *The world of mathematics.* Resonance Publications. www.resonancepub.
    com/mathematics.htm.

QuotationsPage.com (Seneca, A.L.) by Moncur, M. (1994–2007). Quotation Details.
    www.quotationspage.com/quote/29079.html.

Reading academic standards. 2003. Phoenix: Arizona Department of Education.

Reeves, D. B. 2004. *Accountability in action: A blueprint for learning organizations.* 2nd ed.
    Englewood, CO: Advanced Learning Press.

Romance, N., & M. Vitale. 1992. A curriculum strategy that expands time for
    in-depth elementary science instruction by using science-based reading strategies:
    Effects of a year-long study in grade four. *Journal of Research in Science Teaching,*
    29 (6): 545–554.

Silberman, M., & C. Auerbach. 2006. *Active training: A handbook of techniques,
    designs, case examples, and tips.* San Francisco, CA: Jossey-Bass/Pfeiffer.

Tant, C. 1992. *Projects: Making hands-on science easy.* Angleton, TX: Biotech
    Publishing.

U.S. Department of Education, Institute of Education Sciences, National Center for
    Education Statistics. 1996, 2000, 2005. *National assessment of educational progress
    (NAEP).* Washington, D.C.: National Center for Education Statistics.

Walker, H., et al. 1996. Integrated approaches to preventing antisocial behavior
    patterns among school-age children and youth. *Journal of Emotional and
    Behavioral Disorders,* 4 (4): 94–209.

Wilcox, J. 2006. Chicago teachers learn to build academic vocabulary. *ASCD
    Education Update,* 48 (6): 1–2.

Yager, R. E. 1983. The importance of terminology in teaching k–12 science.
    Journal of Research in Science Teaching, 20 (6): 577–588.

Suggested Reading

Abell, S. K., & M. J. Volkman. 2006. *Seamless assessment in science.* Portsmouth, NH: Heinemann.

Bochinski, J. B. 1991. *The complete handbook of science fair projects.* New York: John Wiley and Sons.

Bosak, S. V. 2000. *Science is . . . : A source book of fascinating facts, projects, and activities.* Richmond Hill, Ontario, Canada: Scholastic.

Brandwein, P. F. 1968. *The method of intelligence.* Presentation to NSTA Convention, Toronto.

Campbell, B., & L. Fulton. 2003. *Science notebooks: writing about inquiry.* Portsmouth, NH: Heinemann.

Candler, L. 1995. *Cooperative learning and hands-on science.* San Juan Capistrano, CA: Kagan Cooperative Learning.

Churchill, E. R. 1992. *Amazing science experiments with everyday materials.* New York: Sterling.

Douglas, R., M. P. Klentschy, & K. Worth. 2006. *Linking science and literacy in the k–8 classroom.* Arlington, VA: NSTA Press.

Duke, N. K., & P. D. Pearson. 2002. Effective practices for developing reading comprehension. In A. E. Farstrup and S. J. Samuels,eds.), *What research has to say about reading instruction,* 3rd ed., 205–242. Newark, DE: International Reading Association.

Furtak, E. M. 2009. *Formative assessment for secondary science teachers.* Thousand Oaks, CA: Corwin Press.

Harvey, S. 1998. *Nonfiction matters: Reading, writing and research in grades 3–8.* Portland, ME: Stenhouse Publishers.

Holbrook, A. M. 2006. *Write to know: Nonfiction writing prompts for middle school science.* Englewood, CO: Advanced Learning Press.

Holley, D. 1996. *Sciencewise book 1: Discovering scientific process through problem solving.* Pacific Grove, CA: Critical Thinking Books and Software.

Holley, D. 1996. *Sciencewise book 2: Discovering scientific process through problem solving.* Pacific Grove, CA: Critical Thinking Books and Software.

Holley, D. 1999. *Sciencewise book 3: Discovering scientific process through problem solving*. Pacific Grove, CA: Critical Thinking Books and Software.

Howard, L. 2006. *Ready for anything: Supporting new teachers for success*. Englewood, CO: Advanced Learning Press.

Karplus, R. 1965. *Theoretical background of the science curriculum improvement study*. Berkeley, CA: Science Curriculum Improvement Study.

Keely, P. 2008. *Science formative assessment: 75 practical strategies for linking assessment, instruction and learning*. Thousand Oaks, CA: Corwin Press.

Keeley, P., F. Eberle, & L. Farrin. 2005. *Uncovering student ideas in science: 25 formative assessment probes*, Vol. 1. Arlington, VA: NSTA Press.

Keeley, P., F. Eberle, & L. Farrin. 2005. 2007. *Uncovering student ideas in science: 25 formative assessment probes*, Vol. 2. Arlington, VA: NSTA Press.

Lanatz, Jr., H. B. 2004. *Rubrics for assessing student achievement in science grades k–12*. Thousand Oaks, CA: Corwin Press.

LePatner, M. 2005. *Write to know series: Nonfiction writing prompts for science*. Englewood, CO: Advanced Learning Press.

Liem, T. L. 1992. *Invitations to science inquiry*. Chino Hills, CA: Science Inquiry Enterprises.

Mandell, M. 1990. *Simple experiments with everyday materials*. New York, NY: Sterling.

Marzano, R. 2007. *The art and science of teaching*. Alexandria, VA: ASCD.

Mathematics and Engineering Staff Center for Science. 1998. *Every child a scientist: Achieving science literacy for all*. Washington, D.C.: National Academies Press.

Popham, J. 2003. *Test better, teach better: The instructional role of assessment*. Alexandria, VA: ASCD.

Popham, J. 2007. *Classroom assessment: What teachers need to know*, 5th ed. Boston, MA: Allyn and Bacon.

Rasmussen, G. 1990. *Play by the tules*. Standwood, WA: Tin Man Press.

Reeves, D. B. 2004. *Making standards work*. Englewood, CO: Advanced Learning Press.

Rezba, R. J., C. Sprague, & R. L. Fiel. 2003. *Learning and assessing science process skills*. Dubuque, IA: Kendall/Hunt.

Rhoton, J., & P. Bowers. 2003. *Science teacher retention: Mentoring and renewal*. Arlington, VA: NSTA Press.

Rohrig, B. 1997. *150 Captivating chemistry experiments using household substances*. Plain City, OH: FizzBang Science.

Rohrig, B. 2002. *150 more captivating chemistry experiments using household substances*. Plain City, OH: FizzBang Science.

Stanish, B. 1987. *Inventioneering: Nurturing intellectual talent in the classroom*. Catharge, IL: Good Apple.

Tolman, M. N. 2001. *Discovering elementary science: Method, content, and problem-solving activities*, 3rd ed. Boston, MA: Allyn and Bacon.

U.S. Congress, Office of Technology Assessment. February 1992. *Testing in American schools: Asking the right questions, OTA-SET-519.* Washington, D.C.: U.S. Government Printing Office.

VanCleave, J. 1985. *Janice VanCleave's the human body for every kid: Easy activities that make learning science run.* New York: John Wiley and Sons.

VanCleave, J. 1989. *Janice VanCleave's chemistry for every kid: 101 easy experiments that really work.* New York: John Wiley and Sons.

VanCleave, J. 1990. *Janice VanCleave's biology for every kid: 101 easy experiments that really work.* New York: John Wiley and Sons.

VanCleave, J. 1991. *Janice VanCleave's astronomy for every kid: 101 easy experiments that really work.* New York: John Wiley and Sons.

VanCleave, J. 1991. *Janice VanCleave's earth science for every kid: 101 easy experiments that really work.* New York: John Wiley and Sons.

VanCleave, J. 1991. *Janice VanCleave's physics for every kid: 101 easy experiments in motion, heat, light, machines, and sound.* New York: John Wiley and Sons.

VanCleave, J. 1995. *Janice VanCleave's rocks and minerals: Mind-boggling experiments you can turn into science fair projects.* New York: John Wiley and Sons.

VanCleave, J. 1995. *Janice VanCleave's weather: Mind-boggling experiments you can turn into science fair projects.* New York: John Wiley and Sons.

VanCleave, J. 2000. *Janice VanCleave's guide to more of the best science fair projects.* New York: John Wiley and Sons.

Vecchione, G. 1998. *100 first-prize make-it-yourself science fair projects.* New York: Sterling.

Wollard, K. 1993. *How come?* New York: Workman.

Yager, R. E. 2005. *Exemplary science: Best practices in professional development.* Arlington, VA: NSTA.

Yager, R. E. 2006. *Exemplary science in grades 5–8: Standards-based success stories.* Arlington, VA: NSTA.

Webography

Resources for Science

American Association of Physics Teachers
  www.aapt.org

American Chemical Society
  www.acs.org

Bill Nye
  www.billnye.com

Biology Corner
  www.biologycorner.com/worksheets/labreport.html

BrainPOP
  www.brainpop.com

Discovery Channel
  www.discovery.com

Do Science
  www.doscience.com

Environmental Protection Agency
  www.epa.gov

Exploratorium: Ten Cool Sites
  apps.exploratorium.edu/10cool/index.php

Extreme Science
  www.extremescience.com

Flinn Scientific
  www.flinnsci.com
  Go to the New Teacher section to get free lab activities via e-mail

Funology
  www.funology.com

How Stuff Works
  www.howstuffworks.com

Internet4Classrooms
  www.internet4classrooms.com/science_elem.htm

Laboratory Safety Institute
  www.labsafety.org

Live Science
>   www.livescience.com

NASA
>   www.nasa.gov

National Association of Biology Teachers
>   www.nabt.org

National Earth Science Teachers Association
>   www.nestanet.org

National Oceanic and Atmospheric Administration
>   www.noaa.gov

National Science Education Standards
>   www.nap.edu

National Science Teachers Association
>   www.nsta.org

National Weather Service
>   www.nws.noaa.gov

Nature
>   www.nature.com

Popular Science
>   www.popsci.com

Red Orbit
>   www.redorbit.com

Science Daily
>   www.sciencedaily.com

The Science Explorer
>   www.exploratorium.edu/science_explorer

Scientific American
>   www.scientificamerican.com

Space
>   www.space.com

Steve Spangler Science
>   www.SteveSpanglerScience.com

Weather Underground
>   www.wunderground.com

Yahoo Science Directory
>   dir.yahoo.com/Science

Resources for Conceptual Understanding

Enhancing the Conceptual Understanding of Science
>   www.pilambda.org/horizons/v81-2/gabel.pdf

Resources for Reading and Writing

International Reading Association
    www.reading.org

Lone Star Learning
    www.lonestarlearning.com

Science Writing Prompts
    www.ebecri.org/custom/homepagewritingprompts.html

Video Writing Prompts
    www.teachersdomain.org/resource/wlvt07-scitech.skimotion/

Resources for Problem Solving

Access Excellence at the National Health Museum
    www.accessexcellence.org

Brainers: Brain Teasers from PedagoNet
    www.pedagonet.com/brain/brainers.html

Hunkin's Experiments (over 200 home experiments)
    www.HunkinsExperiments.com

Kids Science Experiments
    www.kids-science-experiments.com

PBS Kids
    www.pbskids.org

PBS Teachers
    www.pbs.org/teachers/sciencetech/

Rubistar
    Rubistar.4teachers.org

Science Experiments You Can Do at Home or School
    www.sciencebob.com/experiments/index

Science is Fun in the Lab of Shakhashiri
    scifun.chem.wisc.edu

TryScience.org
    www.tryscience.org

Resources for Common Formative Assessments

Exemplars
    www.exemplars.com/materials/samples/science

Whelmers: McREL's Accessible Science Series
    www.mcrel.org/resources/whelmers/index.asp

IPL Kidspace: Science Fair Project Resource Guide
    www.isd77.k12.mn.us/resources/cf/SciProjInter.html

Performance Assessment Links in Science
    pals.sri.com

Index

"Conceptual Understanding of Science" (Gabel), 102
Conceptual units, 35, 62, 74, 110 (exh.), 111-115 (exh.), 128, 131, 141, 145, 150, 158, 160, 164, 206, 209, 212, 214, 227, 228; designing, 101, 103-104, 106-109, 183, 229; performance tasks and, 190; rationale for, 102-103; teaching, 183; template for, 104, 266-270 (exh.)
Connections, 19, 20, 21 (exh.), 41, 101; efficacy and, 155; making, 61, 124, 162
Consistency, 4, 19, 20, 21 (exh.), 41
Content, 1, 4, 20, 158, 159, 162, 200, 208, 216, 228, 229; conceptual understanding of, 155; exploring, 51; mapping out, 229; process skills and, 32; recall of, 168; template for, 145, 146 (exh.), 149; understanding, 10, 131
Creative Challenge Activities, 70, 256-265 (exh.)
Creative Connector, 124, 274 (exh.)
Cues, verbal/nonverbal, 34, 124
Current Characteristics of the Science Program in the School or District, 234-238 (exh.)
Curriculum, 4, 11, 74, 228, 229, 230, 231; standards and, 2, 9, 42
Curriculum and Evaluation Standards for School Mathematics (National Council of Mathematics), 3

Data, 37; collecting, 9, 181, 190
Data sheets, 63, 72, 74, 161, 200, 209, 216; problem-solving, 62, 255 (exh.)
Data Teams, 103, 181, 182, 183, 184
Demonstrations, 26, 34, 76, 119
Details, 136, 176; graphic organizer for, 171 (exh.)
Diagrams, 161, 168
Differentiation, 42-43, 44 (exh.), 164, 182, 190, 195; strategies for, 45-46 (exh.)
Dimensions of Thinking: A Framework for Curriculum and Instruction (Marzano), 116
Discrepant events, 74, 77-78 (exh.), 108, 228-229; competition, 79, 79 (exh.); templates for, 80 (exh.), 81-98 (exh.); things to remember doing, 75-76
Discussion Director, 124, 275 (exh.)

Discussions, 35, 123, 124
Diversity, 199, 207, 215
Drawing conclusions, graphic organizer for, 172 (exh.)
During reading, 131, 136-139

Effectiveness, 19-21 43, 47, 50, 109, 123, 130, 136, 139, 144, 164, 167, 179, 180 (exh.), 229
Einstein, Albert, 53, 69
Equipment, 227; emergency, 27, 34; safety/procedures with, 199, 207, 215
Equipment Bingo, 24-25, 253 (exh.)
Essential Questions, 7, 10, 17, 41, 49, 101, 107, 108, 143, 145, 179, 180, 183
Every Child a Scientist: Achieving Scientific Literacy for All, 18
Everybody Counts: A Report to the Nation on the Future of Mathematics Education (National Research Council), 3

Experiments, 26, 34, 70, 158, 159, 176

Facts, science and, 144
Feedback, 70, 74, 162, 181, 182, 193, 195, 231; group, 124; monitoring/providing, 12, 109, 123; regular/timely, 179; from scoring guides, 71
Five Easy Steps, 7-8, 103, 181; Balanced Science Alignment, 185 (exh.); implementation of, 10, 225, 226, 233, 239; strategic planning for, 240-244 (exh.); strategies from, 131-132, 134, 136; time frame for/year 1: 245 (exh.); time frame for/year 2: 246 (exh.); time frame for/year 3: 247 (exh.)
Five-Three-One, 136, 138 (exh.)
Flinn Scientific, 26
Focus Board, 41-42, 41 (exh.)
Four-Square Thinking, 139, 140 (exh.)
Functions, 4, 209, 212

Gabel, Dorothy, 102
Gems of Wisdom, 160, 161
Goals, 3, 20, 102, 122, 181, 209, 217
Graphic organizers, 104, 127, 132, 164; examples of, 133 (exh.), 169 (exh.), 170 (exh.), 171 (exh.), 172 (exh.), 173 (exh.), 174 (exh.), 175 (exh.), 205 (exh.), 213 (exh.), 221 (exh.); using, 106, 168, 176, 204, 212, 220

Homework, 42, 108, 229
How People Learn (Bransford, Brown, and Cocking), 144
Hubble, Edwin Powel, 61
Hypotheses, 62, 109, 120, 209

Illustrations, 61, 164
Illustrator, 124, 277 (exh.)
Implementation, 10, 12 (exh.), 103, 182, 183, 184, 225; framework for, 233, 239; questions about, 226-231, 233; time frame for, 151
Independence, 55, 199, 207, 215
Inference, 52, 53, 75, 200, 208, 216
Information, 123, 132, 136, 139, 160, 184, 228; abundance of, 129; acquiring, 43, 134; comprehension of, 128-129; evaluating, 168; factual, 187; mastery of, 7, 201, 204, 206, 209, 212, 214, 217, 220, 222, 230, 233, 249, 280-282; organizing/synthesizing, 122; recall of, 168; recording, 75, 76; science, 159, 166, 167; understanding, 127-129
Inquiry, science, 1, 4, 42-43, 50-52, 51-55, 74, 117
Inquiry and the National Science Education Standards (National Research Council), 144
Instruction, 102, 107, 158, 180, 233, 239; assessments and, 9; components of, 21 (exh.); effectiveness of, 179; large-/small-group, 200, 208, 216; planned, 104; practices, 101, 143; rethinking, 232 (exh.)
Internet, 31, 79, 129, 130, 150
Interpretations, graphic, 200, 208, 216